KV-268-673

TELECOMMUNICATIONS

Other titles in the Project

Physics   Robert Hutchings
Medical Physics   Martin Hollins
Energy   David Sang and Robert Hutchings
Nuclear Physics   David Sang

Biology   Martin Rowland
Applied Genetics   Geoff Hayward
Applied Ecology   Geoff Hayward
Micro-organisms and Biotechnology   Jane Taylor

UNIVERSITY OF BATH • SCIENCE 16-19

Project Director: J. J. Thompson, CBE

TELECOMMUNICATIONS

JOHN ALLEN

Nelson

Thomas Nelson and Sons Ltd
Nelson House, Mayfield Road
Walton-on-Thames, Surrey
KT12 5PL, UK

51 York Place
Edinburgh
EH1 3JD, UK

Thomas Nelson (Hong Kong) Ltd
Toppan Building 10/F
22a Westlands Road
Quarry Bay, Hong Kong

Thomas Nelson Australia
102 Dodds Street
South Melbourne
Victoria 3205 Australia

Nelson Canada
1120 Birchmount Road
Scarborough Ontario
M1K 5G4 Canada

© John Allen 1990

First published by Macmillan Education Ltd 1990
ISBN 0-333-47627-1

This edition published by Thomas Nelson and Sons Ltd 1992

ISBN 0-17-448216-7
NPN 9 8 7 6 5 4 3 2

Printed in Hong Kong.

Contents

The Project: an introduction

The **University of Bath · Science 16–19 Project**, grew out of a reappraisal of how far sixth form science had travelled during a period of unprecedented curriculum reform and an attempt to evaluate future development. Changes were occuring both within the constitution of 16-19 syllabuses themselves and as a result of external pressures from 16+ and below: syllabus redefinition (starting with the common cores), the introduction of AS-level and its academic recognition, the originally optimistic outcome of the Higginson enquiry; new emphasis on skills and processes, and the balance of continuous and final assessment at GCSE level.

This activity offered fertile ground for the School of Education at the University of Bath to join forces with a team of science teachers, drawn from a wide spectrum of educational experience, to create a flexible curriculum model and then develop resources to fit it. This group addressed the task of satisfying these requirements: these requirements:

- the new syllabus and examination demands of A- and AS-level courses;
- the provision of materials suitable for both the core and options parts of syllabuses;
- the striking of an appropriate balance of opportunities for students to acquire knowledge and understanding, develop skills and concepts, and to appreciate the applications and implications of science;
- the encouragement of a degree of independent learning through highly interactive texts;
- the satisfaction of the needs of a wide ability range of students at this level.

Some of these objectives were easier to achieve than others. Relationships to still evolving syllabuses demand the most rigorous analysis and a sense of vision – and optimism – regarding their eventual destination. Original assumptions about AS-level, for example, as a distinct though complementary sibling to A-level, needed to be revised.

The Project, though, always regarded itself as more than a provider of materials, important as this is, and concerned itself equally with the process of provision – how material can best be written and shaped to meet the requirements of the educational market-place. This aim found expression in two principal forms: the idea of secondment at the University and the extensive trialling of early material in schools and colleges.

Most authors enjoyed a period of secondment from teaching, which not only allowed them to reflect and write more strategically (and, particularly so, in a supportive academic environment) but, equally, to engage with each other in wrestling with the issues in question.

The Project saw in the trialling a crucial test for the acceptance of its ideas and their execution. Over one hundred institutions and one thousand students participated, and responses were invited from teachers and pupils alike. The reactions generally confirmed the soundness of the model and allowed for more scrupulous textual housekeeping, as details of confusion, ambiguity or plain misunderstanding were revised and reordered.

The test of all teaching must be in the quality of the learning, and the proof of these resources will be in the understanding and ease of accessibility which they generate. The Project, ultimately, is both a collection of materials and a message of faith in the science curriculum of the future.

J.J. Thompson
January 1990

How to use this book

The main aim of this book is to describe the essential science behind telecommunication systems. However there is a second aim, which is to show the wider context of the impact of telecommunications on society. The book is particularly geared to the requirements of students who are following an A-level or AS-level course in Physics and are taking an option in telecommunications. Telecommunications involves a rapidly changing technology. The basic principles may not change very much but the implementation of these principles will change considerably over the next few years.

The book covers the range of topics usually covered in such an option (except for a few items that will normally be covered in an A-level Physics textbook) and is divided into three themes. The first theme (Chapters 1 and 2) describes the history of the various telecommunication systems and the background theory required for understanding the subsequent chapters. The second theme (Chapters 3 and 4) describes the use of radio waves in telecommunications. The final theme (Chapters 5 to 7) describes the more recently introduced satellite and optical fibre systems. It looks at what is happening in telecommunications today and what will probably happen in the near future.

The seven chapters are very similar in layout and further details are given opposite. Chapters 1 and 2 should be studied first and chapter 7 last. Chapter 2 is probably the most important, as the understanding of it is essential for studying the rest of the book. It is not very likely that you will completely understand this chapter at the first reading. It will be necessary to come back to sections of it when reading the following chapters. The order of studying Chapters 3 (and then 4), 5 or 6 is up to you. When working through the chapters it would be useful to have the syllabus you are using at hand as there may be sections in the book you may not need to study, though reading them will help broaden your understanding.

I would like to express my thanks to those people at British Telecom, especially Geoff Groom, who have given me a great deal of help in writing this book. However any errors that are left are entirely my responsibility.

Prerequisites

These are presented at the beginning of each theme to remind you of the scientific content which will help you to fully understand the material in the theme. There are not many of these but, if you have not covered any of them in your studies, you will need to consult a suitable text or your teacher.

Learning Objectives

Each chapter starts with a list of learning objectives which outline what you should gain from studying the chapter. These statements often link closely to statements in the course syllabus and can be used to help you make notes for revision, as well as checking your progress.

Questions

There are two types of questions. There are in-text questions at the end of each section (except for Chapter 7) that are designed to test your understanding of the section and can be answered from material presented in that section, though a few questions may require additional thought and information. You should do these questions as you come to them. You may find some of these questions rather difficult and you may have to get help from your teacher. At the ends of Chapters 1 to 6 you will find questions that can cover any area of the chapter (or previous chapters). A large number of these are taken from past examination papers. A calculator with logarithmic functions is necessary for some of the questions. Numerical answers are given at the back of the book.

Investigations

There are a number of investigations in the book. Many of these are designed to be open-ended in their approach. Often there will be no single 'correct' answer. In the Resources section there may be further details on the apparatus required or extra materials that could be used.

Comprehension

These are designed to test your understanding of a piece of scientific text related to telecommunications.

Margin Notes

These are pieces of information that may be anecdotal or interesting.

Summaries

Each chapter finishes with a brief summary of its contents. These summaries, together with the learning objectives, should give you a clear overview of the material and allow you to check your own progress.

Further Reading and Resources

The Further Reading section contains reference to books that are useful throughout the course or for a particular chapter. The Resources section gives further details, where necessary, about the equipment needed for the investigations and any extra materials that would be useful.

Index

This lists important topics, concepts and terms that it is important for you to know and understand.

Acknowledgements

The author and publishers wish to thank the following who have kindly given permission for the use of copyright material:

Oxford and Cambridge Schools Examination Board, University of Cambridge Local Examinations Syndicate and University of London School Examinations Board for questions from past examination papers. The National Museum of Science and Industry for Figs 1.16 and 1.17 from *Submarine Telegraphy: The Grand Victorian Technology* by Bernard S. Finn.

The author and publishers wish to acknowledge, with thanks, the following photographic sources:

AT&T Bell Telephones *p 6 top;* European Space Agency *p 67 bottom;* ICOM UK *p 31;* Illustrated London News Picture Library *pp 7, 9 centre and bottom;* The Institute of Electrical Engineers *pp 2, 32;* Marconi *p 41;* Mary Evans Picture Library *p 35 right;* David Neal *pp 18, 35 left, 40, 48, 57, 59, 73, 80, 81, 93, 94;* Panasonic *p 102 top;* TASS *p 69;* Telecom Technology Showcase *p 6 bottom;* Telefocus *pp 11, 12 top and bottom, 63, 67 left, 68, 86, 88, 93, 101, 102 bottom, 103, 104, 105;* Trustees of the Science Museum, London *pp 1, 3, 4, 9 top, 31, 48;* ZEFA Picture Library *p 67 right.*

Every effort has been made to trace all the copyright holders, but if any have been inadvertently overlooked the publishers will be pleased to make the necessary arrangement at the first opportunity.

Theme 1

INTRODUCTION TO TELECOMMUNICATIONS

This theme consists of two quite different chapters. The first deals with the history of the various telecommunication systems. The discovery of the magnetic effect of an electric current led to the invention of the telegraph and then the telephone. The discovery of suitable electromagnetic waves led to radio and, later, television. In the 1960s and 70s two new systems came into existence – communication using satellites and transmission by optical fibres. These two have brought many important changes into telecommunications. The second chapter deals with the basic ideas and principles required for an understanding of modern telecommunication systems.

PREREQUISITES

Before you study this theme you should have some familiarity with the nature and properties of waves.

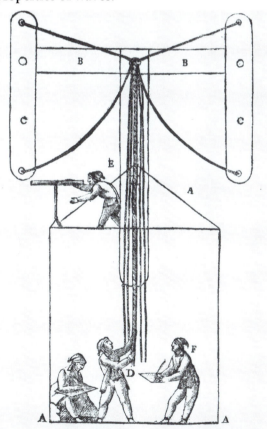

EXPLANATION OF THE TELEGRAPH.

A A A.—A Tent erected at the distance of every four leagues.

B B.—A great Cross-jack or Beam, 16 feet long.

C C.—Small Cross-jacks. N. B. These three Jacks B B. C C. produce between them 280 movements, by means of ropes which concentrate in the tent : each movement lasts 20 seconds ; that is to say, the machine is stopped, and remains motionless in a notch during 20 seconds, and then it begins the second movement, and thus goes on with the rest.

D.—A man who pulls the rope, and directs the movement of the Machine.

E.—A man with a telescope who observes the next Machine and points out to D the movements which he is to imitate.

F F.—Two men writing the names of each movement, which are arbitrary, and compose a cypher of 280 signs, which may be changed at pleasure, and of which the two correspondents field alone the key ; so that all those who work the Telegraph do not know what news they transmit. The great jack only makes four movements, the perpendicular, the horizontal, the first oblique, and the second oblique: the rest have a greater number, and 240 of them are very clear, so that the movement of the jack is, each time, 45 degrees : that is to say, one-eighth of the Circle.

An early military semaphore system. This appeared in a daily newspaper in 1794.

Chapter 1

A HISTORY OF TELECOMMUNICATIONS

LEARNING OBJECTIVES

After studying this chapter, you should be able to:

1. recall the historical development of the following telecommunication systems: telegraph, telephone and radio;

2. know that telecommunication systems use either electric currents or electromagnetic waves for information transfer;

3. recall the initial development of satellite and optical fibre telecommunication systems;

4. understand the effects on society of the various developments in telecommunications.

1.1 THE TELEGRAPH

Fig 1.1 **(a)** Alexander Volta who produced the first battery.

(b) Hans Christian Oersted who, in 1819, discovered the link between electricty and magnetism.

At the beginning of the nineteenth century ways of sending messages and conveying information were very limited. The very big towns had local newspapers and were linked by stagecoaches. Postal services had developed greatly in the eighteenth century. However, for the majority of people the only form of information transfer was by word of mouth, the main sources of information being the passing traveller or the wandering pedlar. The most efficient and fastest means of information transfer were designed for military use. One of these was the mechanical semaphore system used between London and Portsmouth. This used a set of six shutters and the position of the shutters indicated the particular letter being sent. The signals were relayed between towers situated on the tops of hills. At its fastest, this system could send a message between the two places in just three minutes. However, it was useless at night or in foggy conditions.

The magnetic effect of an electric current

The production of a steady electric current from the first cell by Alessandro Volta in 1800 was the first step in the development of a practical electrical telecommunication system. The cell consisted of two different metals (silver and zinc) separated by sodium chloride solution. The metals were arranged in series and the device was called a voltaic pile. In 1819 Hans Christian Oersted discovered the magnetic effect produced by an electric current. When a current flows through a conductor a magnet placed nearby is deflected (see Fig 1.2).

Oersted's discovery led to the invention of the electromagnet by William Sturgeon in 1825. If wire is coiled round a piece of soft iron then the soft iron becomes a temporary magnet when a current passes through the wire. The magnetic force produced by the current is magnified. It was also known that the speed at which an electric current travelled along a wire was incredibly high.

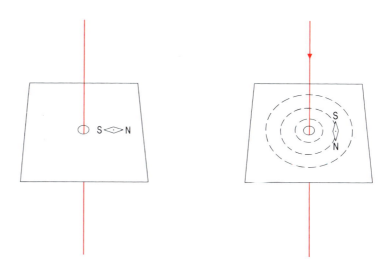

Fig 1.2 The magnetic effect of an electric current.

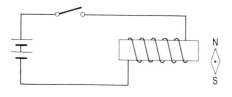

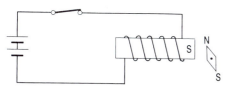

Fig 1.3 Magnifying the magnetic effect of an electric current.

Fig 1.4 An early form of Cooke and Wheatstone's telegraph.

The first telegraph

The first practical electrical **telegraph** ('tele' means at a distance and 'graph' means writing) was developed by William Cooke and Charles Wheatstone. It started operation in 1837 on the railway line between Euston and Camden Town (a distance of about 1.5 km). It was used for signalling the presence of trains going through the single line tunnel between the two stations. Although apparently successful it did not lead to any further use on this railway line, but a more sophisticated version was installed two years later on another railway line between Paddington and West Drayton. Besides its primary use for signalling the presence of trains, the public were allowed to send messages for 1 shilling (5 pence). The system caught the public imagination when, in 1845, a suspected murderer boarded a train at Slough for Paddington, where a telegraphed message had the police waiting for him when he arrived there. He was subsequently found guilty and hanged.

One version of this type of telegraph is shown in Fig 1.4. There were five magnetic needles which could be moved in either direction by electromagnets. By pressing the two appropriate keys at the transmitting end of the line the needles at the receiving end would point to the required letter. The fact that only twenty letters can be transmitted led to some licence with the spelling. The five needle telegraph was soon replaced by devices that were simpler to install but harder to operate.

By 1850 there was about 6000 kilometres of aerial telegraph line in use in the United Kingdom. As with other telecommunication systems developed later, all the telegraph companies were eventually nationalised. This took place in 1868 and the system was then operated by the Post Office. At this time the complete network consisted of about 100 000 km of line and 2800 telegraph offices.

Samuel Morse and his code

In the United States of America a different type of telegraph system evolved. This was conceived by Samuel Morse in 1835. The basic system consisted of a complete electrical circuit with a key at the transmitting end and an electromagnet connected to a device for making marks on paper at the receiving end. The key was operated to make the current flow in the circuit and thus operate the electromagnet. Long or short 'taps' on the key produced long or short marks on the paper. The code used came to be known as the **Morse code**. It should be noted that the most frequently used

Under the Special Patronage of Her Majesty

And H. R. H. Prince Albert

GALVANIC AND MAGNETO

ELECTRIC TELEGRAPH,

GT. WESTERN RAILWAY.

The Public are respectfully informed that this interesting & most extraordinary Apparatus, by which upwards of 50 SIGNALS can be transmitted to a Distance of 280,000 MILES in ONE MINUTE,

May be seen in operation, daily, (Sundays excepted,) from 9 till 8, at the

Telegraph Office, Paddington,

AND TELEGRAPH COTTAGE, SLOUGH.

ADMISSION 1*s.*

" *This Exhibition is well worthy a visit from all who love to see the wonders of science.*"—Morning Post.

Despatches instantaneously sent to and fro with the most confiding secrecy. Post Horses and Conveyances of every description may be ordered by the Electric Telegraph, to be in readiness on the arrival of a Train, at either Paddington or Slough Station.

The Terms for sending a Despatch, ordering Post Horses, &c., only One Shilling.

N.B. Messengers in constant attendance, so that communications received by Telegraph, would be forwarded, if required, to any part of London, Windsor, Eton, &c.

THOMAS HOME, *Licensee.*

G. NURTON, Printer, 48, Church Street, Portman Market.

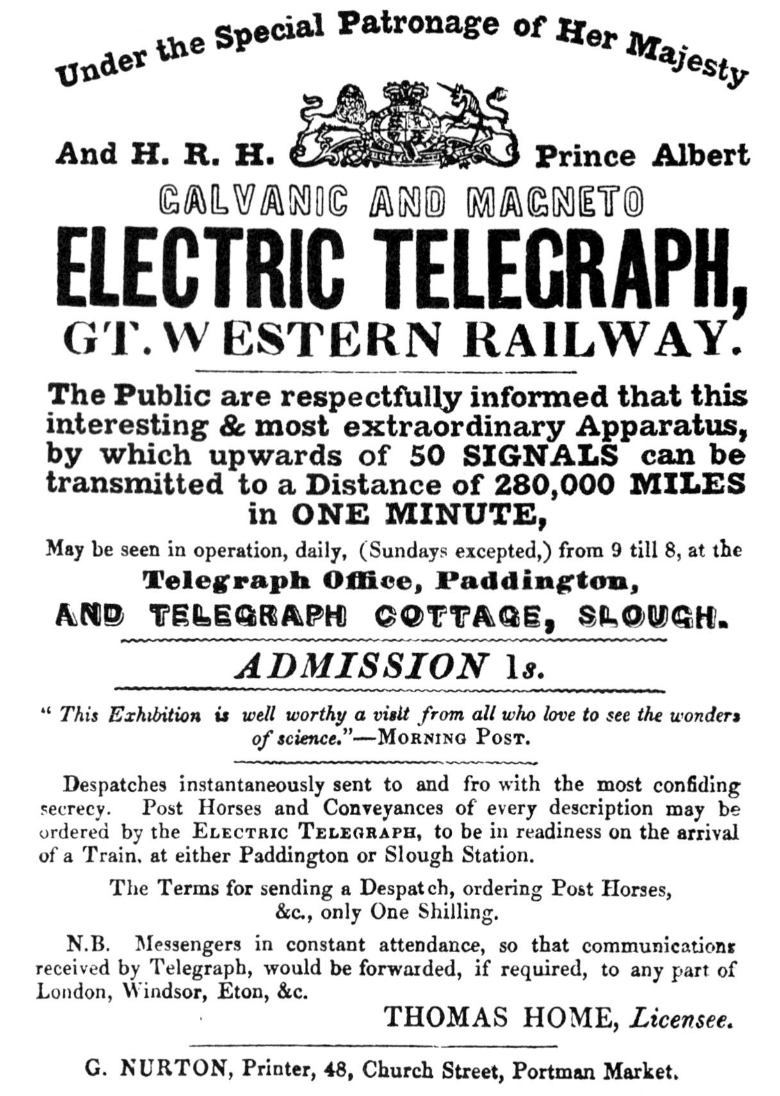

Fig 1.5 An 1845 advertisement describing the marvels of the telegraph.

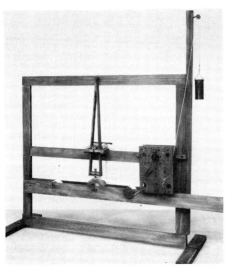

Fig 1.6 **(a)** The first morse receiver. The paper on which the marks were made is in the middle of the picture.

letters of the alphabet have the shortest code and this increases the rate of transmission. However, trained operators could more readily hear the difference between the 'dashes' and 'dots' and this rapidly became the accepted system of message sending. In 1844 the first commercial telegraph in the USA was established between Washington and Baltimore, a distance of about 70 km.

Telegraphy by submarine cable

The usefulness of the telegraph system was soon appreciated and a rapid growth resulted. Eventually the wires of the telegraph system covered the USA and Europe. While this was going on the next step was to extend the telegraph system between countries separated by water. Laying cables under water presented many difficulties not encountered on land. Insulation was vitally important together with the armouring of the cable to prevent damage from such things as trailing anchors. It is important to remember that in those days there was no way of amplifying the signals en route. The first successful underwater cable was laid across the English Channel between England and France in 1851 and, seven years later, a

Morse code

A •—	J •———	S •••
B —•••	K —•—	T —
C —•—•	L •—••	U ••—
D —••	M ——	V •••—
E •	N —•	W •——
F ••—•	O ———	X —••—
G ——•	P •——•	Y —•——
H ••••	Q ——•—	Z ——••
I ••	R •—•	

(b) The morse code.

cable was laid across the Atlantic Ocean. Two ships met in the middle of the Atlantic, each carrying half of the enormous length of cable. The ends were joined together and they sailed off in opposite directions. The cable had only a short operational life before it failed. In 1866 a more permanent link was established. This time the whole cable was carried on one ship, the Great Eastern, built by Isambard Kingdom Brunel.

A network of underwater telegraph cables began to be established around the world (see Fig 1.7). By the beginning of the twentieth century a world-wide system of telegraph cables had been completed with the cable laid across the Pacific Ocean the last link in this chain. Although these telegraph lines required specialist operators and there were limits as to how quickly messages could be transmitted, many were still in use up until the middle of the twentieth century.

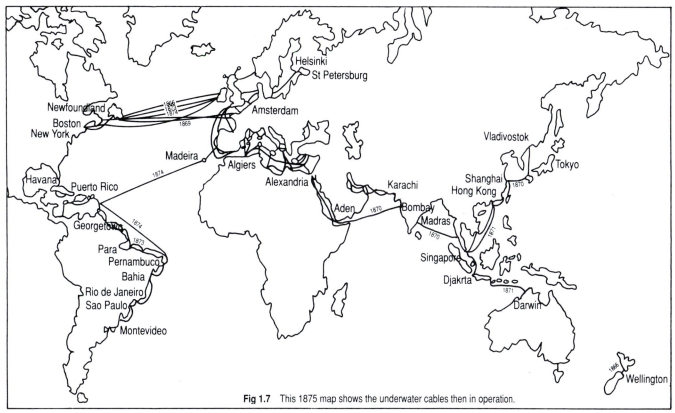

Fig 1.7 This 1875 map shows the underwater cables then in operation.

Further developments

The longer the distance over which the signals were sent the weaker the received signals would be. Consequently there was a limit to the distance between transmitter and receiver. With land telegraphy this presented fewer problems than submarine telegraphy as it was comparatively easy to adjust the transmission distance. Ways were developed of sending more than one message at a time along the wire (this is called multiplexing) and of sending messages simultaneously in both directions. By 1870 messages were entered onto punched tape and then transmitted and received at high speed. The sending of pulses long distances by submarine cable presented difficulties that would not be eliminated until amplifying devices could be built into the cables. One type of telegraph system has survived. This is called telex and started in 1932. Further details about it will be given in Section 7.2.

QUESTION	1.1 Describe the advantages and disadvantages of the telegraph with the previous non-electrical methods of telecommunications.

1.2 THE TELEPHONE

Alexander Graham Bell

The next significant development was the invention of the telephone (the word means speaking at a distance). Alexander Graham Bell was very successful at teaching deaf children to speak and he was engaged in ways of changing speech waves to electrical signals and back again. The complete device consisted of a 'liquid' transmitter and an electromagnetic receiver. A vibrating diaphragm varied the resistance of a liquid. The consequent current fluctuations were transmitted to the receiver where an electromagnet caused another diaphragm to vibrate. In 1876 his invention was patented in the USA. This method of telecommunication allowed people to speak to each other directly without the need for an operator to encode and decode the message. Compared with the telegraph system, the rate at which information could be transferred increased, though it was obviously limited to the speed at which a person could speak.

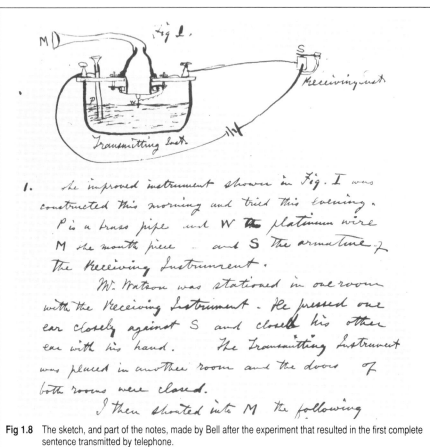

Fig 1.8 The sketch, and part of the notes, made by Bell after the experiment that resulted in the first complete sentence transmitted by telephone.

The following year the transmitter mouthpiece was greatly improved by Thomas Edison. Carbon granules replaced the liquid in Bell's design but the principle of operation was identical. The structure of the telephone mouthpiece remained virtually unaltered for many years. Only recently has its design radically changed, being replaced by moving coil or capacitor microphones. The receiving earpiece has changed remarkably little from Bell's original design.

Telephone exchanges

The first telephone links were installed in large private buildings and used for communication between the occupants. A number of companies were subsequently set up to implement this new technology. Individual subscribers were linked to each other via an exchange. The telephone had

Fig 1.9 This is the first telephone to combine the transmitting and receiving sections (dated 1884).

Almon Strowger was an undertaker who started developing an automatic system when he discovered that the local telephone operator was the wife of a rival undertaker and was diverting many of his business calls to her husband's business.

The first still picture was transmitted by wire in about 1850 by Alexander Bain. The picture was scanned mechanically in a similar way to the TV today and the information was sent as a stream of data.

arrived in the UK soon after the initial demonstrations in the USA and the first exchange was opened in London in 1879 with 10 subscribers. Each subscriber was connected directly to the exchange and the operator had a manual switchboard for connecting the subscribers to each other. In 1888 Almon Strowger invented an automatic telephone selector in the USA that did away with the need for an operator and allowed subscribers to dial their own numbers.

These exchanges were operated by private companies and originally there were few connections between them. In effect the networks of subscribers were very localised. Long distance services were run by the Post Office. In time the situation became chaotic. Eventually the complete system was nationalised in 1912 and run by the Post Office.

After nationalisation of the telephone system there was continuous expansion in the number of exchanges and the areas covered. The first automatic exchange was opened in the UK in 1912 but the Strowger system was not universally adopted until ten years later. The first telephone link between the UK and USA started in 1927 but used radio across the Atlantic. The caller was connected by land line to Rugby. The call was transmitted by radio waves to the USA and then travelled by land line to the recipient. The system allowed only one telephone call at a time and was very expensive. This type of link was further extended to other countries. Until the coming of submarine coaxial telephone cables, this was the only way of making inter-continental telephone calls (even in 1956 there were only 70 inter-continental circuits). At the end of the 1920s it was possible to send pictures along the telephone lines. This is called facsimile transmission or, more usually nowadays, fax. The *New York Times* newspaper was sent to London using this system via the telephone/radio link across the Atlantic. Further details will be given about fax in Section 7.2.

The distance over which telephone transmissions could be sent greatly increased for two reasons. Devices called repeaters were installed at intervals in these long distance cables (the first were installed in 1916).

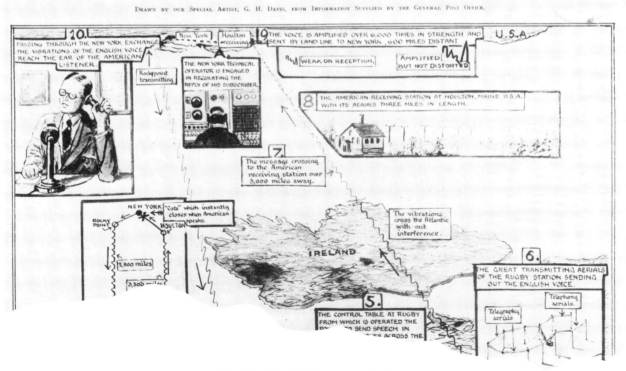

Fig 1.10 On the 8th of January 1927 the *Illustrated London News* described how a telephone call crossed the Atlantic Ocean.

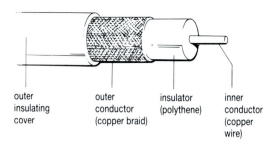

Fig 1.11 A modern coaxial cable.

These detected the incoming signals, amplified them and transmitted them to the next repeater. Also, the reliability of the cables themselves had greatly improved. The first experimental coaxial cable came into service in 1936 and used repeaters housed in buildings constructed along its route. A **coaxial cable** consists of a central thin copper conductor surrounded by an outer cylindrical copper tube or mesh. The two conductors are separated by a solid insulator or air (with spacers to keep the conductors apart). This cable was laid between London and Birmingham and it was possible to send a large number of messages at the same time through it.

QUESTION

1.2 The initial public reaction to the telephone was one of indifference and few people bought the device.
(a) Why do you think this was?
(b) If you had had to market the newly invented telephone what strategies would you have employed?
(c) Compare the advantages and disadvantages of the telegraph and the telephone.

1.3 RADIO AND TELEVISION

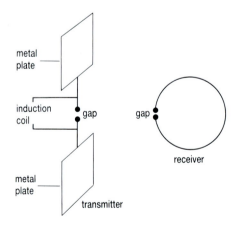

Fig 1.12 A diagrammatic form of Hertz's original apparatus.

Radio

The existence of electromagnetic waves was first postulated by James Maxwell in 1867 in a series of papers which revolutionised electricity and magnetism. He suggested many of the properties that they would have, including their speed. In 1888 Heinrich Hertz successfully transmitted and received electromagnetic waves. These became known as radio waves. A diagrammatic form of his original apparatus is shown in Fig 1.12. The transmitter was an induction coil (this device produces a high voltage from a low voltage battery) that produced a spark across the air gap. The spark consisted of oscillating electric currents and these currents produced the radio waves which travelled away from the gap. At the receiver the electromagnetic waves caused minute sparks to jump across the gap. Although this worked in the laboratory, the receiver was not sensitive enough to detect radio waves that travelled any great distance.

Long distance radio communication

The significance of Hertz's experiments for telecommunication purposes was realised by a number of people but it was Guglielmo Marconi who produced the most important advances. He worked on developing transmitters and receivers to work over longer and longer distances. The process became known as 'wireless' telegraphy. This culminated in the famous 1901 transmission from Cornwall to Newfoundland in Canada, a distance of 3500 km. At this time it was not possible to transmit speech and all messages had to be sent using the Morse code. The transmitter was turned on and off to produce the short and long pulses of radio waves that corresponded to the dots and dashes. Wireless telegraphy allowed mobility

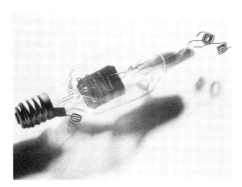

Fig 1.13   A triode made in 1907.

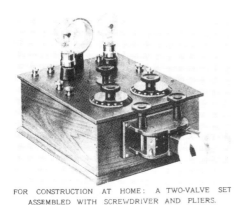

FOR CONSTRUCTION AT HOME: A TWO-VALVE SET
ASSEMBLED WITH SCREWDRIVER AND PLIERS.

Fig 1.14   A home construction radio from 1923.

Fig 1.15   A TV receiver from 1935.

The first outside TV broadcast occurred during the
Coronation procession of George VI in 1937. Three
cameras were set up to show the procession as it
passed Hyde Park Corner, London.

of the transmitter and receiver, unlike the earlier systems.

Ships were the first main users of wireless communications (by 1918 the maximum range was of the order of 500 km and it was estimated that 5000 ships carried a wireless). The murderer Dr Crippen and his accomplice were caught by the use of wireless. They were escaping to the USA and were recognised by the captain of the ship in which they were travelling. He radioed back to London and a police officer boarded a faster ship. He eventually caught up with Crippen's ship and arrested them both. It was not until 1934 that the first radio-telephone link between ship and shore was established.

Among the many inventions and discoveries that improved radio communication, two electronic devices were particularly important. The first was the **diode valve** of John Ambrose Fleming in 1904. This acted in the same way as the semiconductor diode in use today. Its main use in radio receivers was as a 'detector' of the radio waves. It could also be used for changing alternating to direct current (this was very useful as valves required high voltages and the mains supplies could then be used). By adding an extra metal grid to the diode valve, Lee de Forest produced a valve that could amplify the signal. This was called a **triode valve**. This technology soon allowed radio waves to be generated at fixed frequencies. This is essential if there are a number of transmissions taking place simultaneously. Radio had become the first electronic telecommunication system as opposed to the earlier electrical ones.

Transmitting speech

Although the transmission of speech by radio waves is now taken for granted all early transmissions used the Morse code. The first successful attempt to broadcast speech was in 1906 in the USA and the result was heard up to 300 kilometres away. The first commercial broadcasting station started operating in Pennsylvania USA in 1920 and the first station started in the UK two years later. Pressure from potential broadcasters and others resulted in the British Broadcasting Company (later renamed the Corporation) being set up under the auspices of the Post Office. The company had to satisfy a number of technical requirements when broadcasting programmes but received half the money from the receiving licences that people had to buy and royalties on the radio sets sold.

The earliest radio transmissions used low frequency radio waves. Although these could travel a considerable distance high powered transmitters and large aerials were needed. With developments in equipment and much experimentation it was found that higher frequency waves could travel equivalent distances with lower powered transmitters and smaller aerials. By 1930 there was world-wide coverage by radio although the Morse code had to be used on the long distance transmissions. But problems due to atmospherics and solar radiation often caused interference on long distance transmissions and much ingenuity was expended in minimising these effects.

Television

John Logie Baird's invention of the television (the word means seeing at a distance) was first demonstrated in 1926. It was electromechanical (involving both electrical and mechanical sections) and unfortunately this did not lend itself to simple domestic use. Another completely electronic system (using a cathode ray tube for display of the picture) was being developed at the same time. It was decided that the BBC would test both types before a decision was made on which one to adopt. In 1936 test transmissions started using a tower 100 metres high in Alexandra Park, North London. The completely electronic system proved more satisfactory and was adopted for all future broadcasts.

1.4 DEVELOPMENTS SINCE 1945

The Second World War (1939 to 1945) put a stop to all non-military telecommunication development. But much of the work done with military applications in mind had important uses in commercial telecommunications after the war. The most important developments were: radio transmissions using extremely high frequencies (the type of waves used are called microwaves), radar, the first electronic computers and high altitude rockets.

The telephone

From 1956 up until 1968 all submarine cables were laid with underwater repeaters that used valves instead of transistors. The reason was that the designers knew that the proven technology would last the requisite life (25 years) of the cable.

The first repeater in an underwater coaxial telephone cable had been installed during the war between Anglesey and the Isle of Man. This was valve operated but then, in 1948, the invention of the transistor was announced. The **transistor** was the first in a long line of electronic devices made from semiconductor materials. These devices worked at lower voltages, consumed less power and lasted much longer than valves. The transistor started a new era in telecommunications.

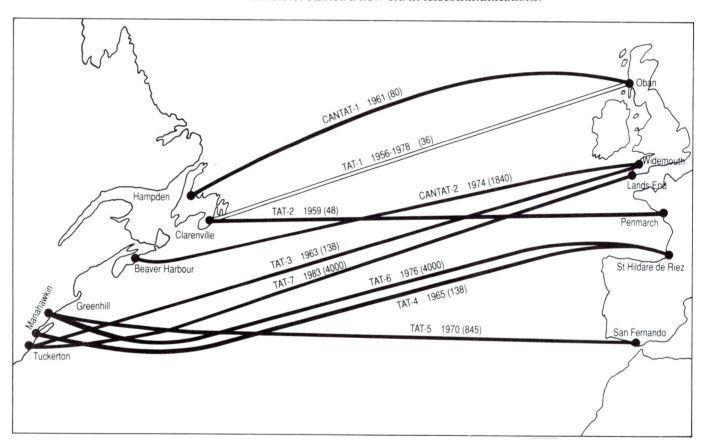

Fig 1.16 This map shows the nine coaxial transatlantic telephone cables with their date of laying and capacities.

The first Transatlantic Telephone cable became operational in 1956 and was called TAT-1. It required a repeater every 60 km and could carry 36 simultaneous telephone conversations. The last transatlantic cable to be laid using the standard coaxial cable techniques is called TAT-7 and can transmit 4000 simultaneous telephone conversations.

International STD started in 1963 when callers in London or Paris could directly call each other. The first intercontinental link was started in 1970 between London and New York. International STD is now called International Direct Dialling (IDD).

In 1958 the first exchange using Subscriber Trunk Dialling (STD) was opened in Bristol. This enabled the subscriber to dial long distance calls without being routed through the operator. The UK entered an era that produced a completely automated system by 1979. Through the 1960s the development effort was on exchanges with mechanical relays involving electronic control. In 1980 the first digital switching exchange controlled by computer was opened. The name given to this type of exchange is System X and it is discussed further in Section 7.1. Nowadays it is possible to dial to nearly two hundred countries from anywhere in the UK. Since 1960 the number of telephone calls has doubled roughly every ten years.

Radio and television

The use of radio for communication purposes has greatly increased. Developments in electronics and equipment have allowed higher and higher frequencies to be used. Higher frequencies allow larger numbers of signals to be transmitted. It had been thought that the coming of telecommunication satellites would reduce the use of radio but this has not happened. On the broadcasting side developments produced great improvements in reception. Very high frequency (VHF) broadcasting allows high quality reception and was started in 1955. The first transistor radios were produced in 1956 allowing broadcasts to be received anywhere with the help of a low voltage battery.

In 1950 there were about 50 000 television sets in the UK. Ten years later, networks covering the UK had been established by both the BBC and the Independent Television Authority and an estimated 10 000 000 sets were in use. The first colour broadcasts in the UK took place in 1967.

The Telecom Tower can now handle nearly 150 TV channels or 2×10^5 telephone conversations. The telephone exchange at its base deals with one quarter of all the calls made in London.

Fig 1.17 Telecom Tower.

The first attempt to transmit TV pictures from the USA to the UK was a failure. The simple reason was that a diode had been placed in the receiving circuitry the wrong way round due to a misunderstanding in the instructions. The following day the diode was reversed and reception was perfect.

Microwave links

As early as 1949 London and Birmingham were linked by a series of microwave transmitters and receivers that were used for sending television pictures between the two cities. Microwaves can be focused into beams using suitable dish aerials (these can be seen near the top of Telecom Tower in Fig 1.17) and these beams are highly directional. The beam from the transmitting aerial travels in a straight line to the receiving aerial. The link was the forerunner of the network that now covers the UK. The focal point of this network is the Telecom Tower in London. This system was first used just for TV transmissions but now forms a major part of the telephone system as well. All the BBC and independent TV broadcasts are routed around the country using the microwave network.

Satellites

In July 1962 Telstar was launched from Cape Kennedy, USA. This was the first satellite to transmit live television signals. The next day viewers in Europe saw the first TV broadcast from the USA. The satellite orbited the Earth in about 158 minutes and its signals could only be received at one point for about 20 minutes at a time and then only if the receiving ground aerial could accurately track the path of the satellite.

The coming of satellites that could be placed in an orbit high enough to appear stationary above the Earth enabled communications to remain operative 24 hours of the day. The first satellite in such an orbit was SYNCOM II, which was launched in 1963. As the technology has developed, successive satellites have been able to handle larger and larger quantities of transmitted information. Besides being used for telephone links, TV and computer data transmission, satellites are being used for many other purposes such as weather forecasting, remote sensing, navigation and military surveillance.

Fig 1.18 TELSTAR was the first telecommunication satellite. It orbited the earth every 158 minutes at a height ranging between 1000 km and 5500 km.

Optical fibres

The most important recent development in telecommunications has been the optical glass fibre. By using light waves, which are of much higher frequency than radio waves, vast amounts of information can be sent along glass fibres thinner than a human hair. One of the biggest problems in the initial development of optical fibres was the production of glass that was pure enough to enable the light waves to travel long distances through it. The first optical fibre system to be part of the UK telephone network started operation in 1977. By the mid-1990s all the coaxial cables in the UK long distance telephone network will be replaced with optical fibres. In December 1988 the first transatlantic optical fibre system (TAT-8) started operation. This is capable of transmitting 20 000 simultaneous telephone conversations.

Present telecommunication technologies

At the moment there are four main technologies involved in long distance telecommunications. These are coaxial cables, radio (including micro-waves), satellites and optical fibres. Coaxial cables use a well-proven technology but there are limitations in the rates of information transfer and

Fig 1.19   INTELSAT 5. This satellite weighed 1000 kg and when deployed in orbit had a 'wingspan' of 16 metres.

they will be superseded in the near future. Radio and microwave links are still developing and are an integral part of the telecommunications network. These are discussed further in Chapters 3 and 4. The use of satellite systems has increased at a very fast rate and these systems are discussed in Chapter 5. The most recent arrival is the optical fibre and its development has increased at an amazing rate. Its potential capacity for information transfer is unparalleled at the moment. Optical fibre systems are discussed in Chapter 6. Chapter 7 will describe the state of telecommunications today and in the near future.

SUMMARY

The invention of the battery and the discovery of the magnetic effect of the electric current revolutionised the exchange of information. The first technology developed was the telegraph. This was followed by the telephone which allowed people to converse directly without the need for elaborate codes. The discovery of electromagnetic waves led to 'wireless' transmissions. When the ability to transmit speech was available, the development of radio and, later, television for entertainment purposes became possible. Improvements in reliability and the transmission distances were continually being made to all systems. The most recent systems to be developed are those using satellites and optical fibres.

QUESTIONS

In Further Reading at the back of this book you will find some references to books that may help you answer these questions.

1.4 The only two human senses that can be used effectively for telecommunication purposes are sight and hearing. Find out the ways in which human beings have tried over the years to improve on their natural abilities in these areas.

1.5 What do you think has been the most significant discovery or invention in the history of telecommunications? Justify your answer.

1.6 The telephone has been described as 'the world's largest machine'. What do you think this statement means and would you agree with it?

1.7 Write a short historical account on one of the following topics.
   (a) Transmitting pictures by wire (facsimile transmission or fax).
   (b) Sending letters by wire (telex) .
   (c) Sending more than one message along a wire at the same time (multiplexing).

Chapter 2

PRINCIPLES OF TELECOMMUNICATIONS

LEARNING OBJECTIVES

After studying this chapter, you should be able to:

1. know that telecommunication systems are made up from similar sub-systems;

2. describe the difference between analogue and digital signals;

3. use the following terms: cycle, wavelength, amplitude, frequency, speed and phase;

4. know that practical waveforms are synthesised from a number of sinusoidal waveforms;

5. describe the principles behind the following types of modulation: amplitude, frequency, pulse code and phase shift keying;

6. describe the principles behind frequency division multiplexing and time division multiplexing;

7. know the effects of attenuation and electrical noise on the propagation of signals;

8. use the decibel power notation in signal transmission calculations.

2.1 TELECOMMUNICATION SYSTEMS

In Chapter 1 the development of a number of telecommunication systems was discussed. The main function of each system was to transfer information as far and as fast as possible. As the technology developed the distances over which information could be transferred and the rate at which it could be sent have continued to increase. This chapter will explain the basic ideas and principles needed to understand the operation of modern telecommunication systems.

Basic telecommunication systems

Although individual telecommunication systems do not appear to have all that much in common they can be shown to be composed of similar **sub-systems**. The actual sub-systems may be very different in construction but perform similar functions.

A non-electrical telecommunication system will contain the sub-systems illustrated in Fig 2.1. The information from the source has first to be encoded into a form suitable for transmission. The encoded signals are then transmitted to the receiver where the signals are decoded to retrieve the original information. For example, in sending messages by using drums, the information has first to be encoded by the sender into a series of

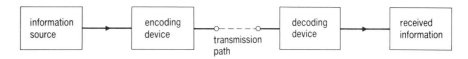

Fig 2.1 A basic telecommunication system.

sounds. The sound waves travel through the air (the transmission path) to the receiver whose ears and then brain decode the sounds to recover the original signal.

The sub-systems in Fig 2.1 can also be applied to simple electrical telecommunication systems but the encoder has to be a transducer. A **transducer** is a device that changes one form of energy into another. In the telephone system, the transducers are the microphone (mouthpiece) and the earpiece. The transmitting microphone changes sound waves into varying electrical currents, which are sent by wires to the receiving earpiece, and this changes the varying currents back into sound waves. In other systems the transmitted signals may be produced from pictures, text or computer data.

Electronic telecommunication systems

With the present electronic systems there are more sub-systems. These are illustrated in Fig 2.2, although a particular system may not use all the sub-systems shown. The information source has first to be changed into electrical signals and then encoded into a form suitable for transmission. The

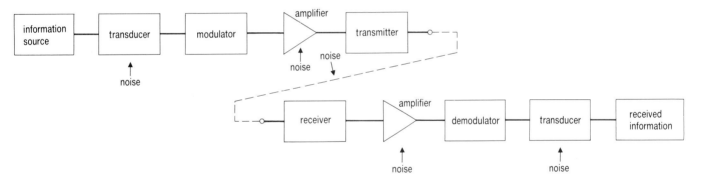

Fig 2.2 An electronic telecommunication system.

process is called modulation and is described in Section 2.4. Many systems need a **transmitter** for getting the signal into the transmission path and a **receiver** at the other end to extract it. The transmission path could be a wire, coaxial cable, air or an optical fibre. The received signal is demodulated and then passed through the appropriate transducer to recover the original information.

You will notice that amplifiers have been added to the system (although not shown in the diagram, amplifiers are also often used in the transmission path). An **amplifier** is a device that will increase the magnitude of a signal. As a signal is transmitted through a system there are two main effects:

*   Energy is lost to the surroundings and so the power carried by the signal decreases. This loss of signal power is called **attenuation** and its effects are discussed in Section 2.6.

*   Noise is added to the signal and this can produce distortion of the signal when it is demodulated. Noise has a special meaning with reference to electrical and electronic circuits. **Noise** refers to any ran-

dom electrical energy that is added to the signal being transmitted. You have probably experienced the effects due to noise when listening to radio or watching television. Types of noise are described in Section 2.6.

Fig 2.2 shows a one-way transmission. A one-way transmission path is called a **channel**. A system that allows two-way transmission is called a **circuit**. A radio broadcast such as Radio 2 is one-way transmission and so requires only one channel. An aeroplane in radio communication with an airport requires two channels i.e. a circuit. With certain systems the signals can be sent along the same transmission path in both directions simultaneously.

QUESTIONS

2.1 A simple light telegraph is illustrated in Fig 2.3(a).
   (a) Using the terminology mentioned in Fig 2.1, describe the system.
   (b) What disadvantages does this type of system have compared with modern telecommunication systems?

2.2 Fig 2.3(b) illustrates a one-way telephone system.
   (a) Explain how the system works.
   (b) Suggest improvements to the system so that it could be used to provide two-way communication.

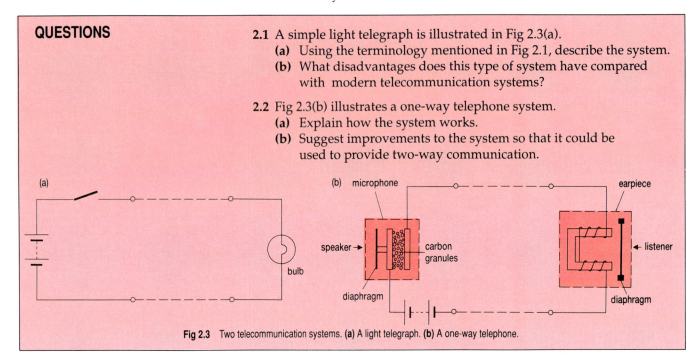

Fig 2.3 Two telecommunication systems. **(a)** A light telegraph. **(b)** A one-way telephone.

2.2 TYPES OF SIGNAL

Analogue and digital signals

These two types of signal are widely used in telecommunications. An **analogue** quantity is one that can assume any value between a minimum and maximum value. Analogue signals vary continuously with time. Examples of telecommunication systems using analogue signals are: local telephone networks and the majority of radio and microwave transmissions (at the moment – this is changing). Analogue signals are

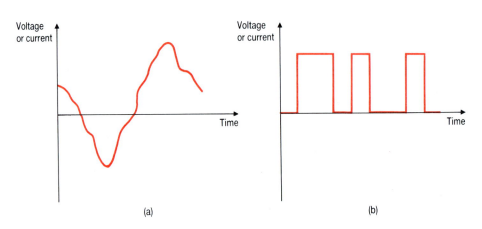

Fig 2.4 **(a)** An analogue signal. **(b)** A digital signal.

PRINCIPLES OF TELECOMMUNICATIONS

attenuated as they travel through a system and require amplification at intervals. The amplifier used in an analogue system is called a **repeater**. It consists of a receiver, an amplifier and a transmitter.

A **digital** quantity can have only fixed values. In telecommunication systems there are usually two values. Digital signals consist of a train of pulses. Examples of telecommunication systems using digital signals are: an increasing number of satellite communication links and most optical fibre transmissions. The trend is towards digital signals and eventually all telecommunication systems will be completely digital. As digital pulses travel through the system they become smaller and distorted. Unlike analogue signals the pulses are actually reproduced in a regenerator. A **regenerator** consists of a receiver, a device for regenerating the pulses and a transmitter. The differing actions of the repeater and regenerator are illustrated in Fig 2.5. Note that, although the signal produced by the repeater is amplified, the effects of the noise are amplified as well. With the regenerator all traces of the noise are removed.

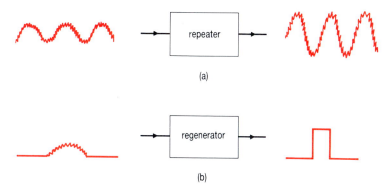

Fig 2.5 **(a)** The repeater. **(b)** The regenerator.

Many telecommunication systems use analogue and digital signals at different stages. For instance, a long distance telephone call will start and end its journey as analogue signals. If it passes through an optical fibre link the signals will be changed into digital ones. It is obviously necessary to be able to change the signals from one type to the other. The electronic devices to do this are called analogue-to-digital converters (**ADC**) and digital-to-analogue converters (**DAC**). With two-way systems using the same transmission path a device called a CODEC (COder/ DECoder) is used.

Binary notation

We are used to expressing numbers in the decimal notation (base 10) and probably never think what the value of each number represents. Each number is ten times the value of the one on its right. For example, the number 198 means:

$$1 \times 10^2 + 9 \times 10^1 + 8 \times 10^0 = 100 + 90 + 8 \text{ (Note: } 10^0 = 1)$$

Binary notation is expressed to base 2. This means that going from right to left, each digit has twice the value of the one on its right. Each binary digit is usually called a **bit** (binary digit). For example, the binary number 1011 means:

$$1 \times 2^3 + 0 \times 2^2 + 1 \times 2^1 + 1 \times 2^0$$

In decimal notation this equals 8 + 0 + 2 + 1 or 11.

ASCII code

Computers use digital signals and this makes them ideal for digital telecommunication purposes. The commonest code is the **ASCII** (American

Standard Code for Information Interchange) **code**. The term is pronounced 'askey'. All the letters, decimal digits, punctuation marks and other symbols are expressed by means of a seven bit code. For instance 'A' is represented by 1000001 (decimal 65). The maximum number of characters that can be represented using a seven bit code is 128.

2.3 CARRYING THE INFORMATION

Characteristics of waves

Most forms of telecommunication involve waves. To understand waves it is perhaps simplest to start with an example – the water waves on the surface of a pond. If a stone is dropped into a pond circular waves spread out from the point where the stone hits the water. While the wave itself spreads out

Direction of motion of the water particles

Direction of wave motion

Fig 2.6 **(a)** A transverse wave. **(b)** Transverse wave motion.

the vibration of the particles is vertically up and down. This can be observed by watching the motion of a cork on the water's surface. This type of wave is called **transverse** and it means that the particles move at right angles to the direction in which the wave is travelling. Waves always transfer energy from one point to another. Most waves require a medium for their transmission. But there is an exception to this – electromagnetic waves, which can pass quite easily through a vacuum.

Parts of the wave

The following terms are used in connection with waves. The waveform shown in Fig 2.7 is sinusoidal as this is a very common type. However, the definitions apply to other types and shapes of waves.

Cycle: one complete oscillation of the wave.

Wavelength: the distance occupied by one complete cycle, e.g. the distance from one crest to the next. The symbol is λ and the unit is the metre.

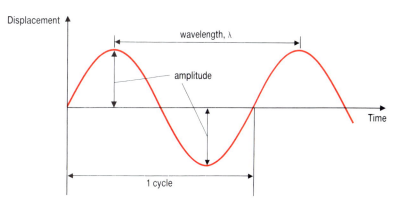

Fig 2.7 Parts of a wave.

Amplitude: this refers to the maximum positive or negative displacement during the cycle.

Frequency: the number of cycles of the wave produced per second. The symbol is f and the unit is the hertz (Hz). The **period** (unit: second) is the duration of one cycle and is equal to $1/f$.

Speed: this is the distance travelled by the wave in one second. The symbol is v and the unit is metres per second. It can be shown that the speed is given by the following expression:

$$v = f\lambda \text{ where } v = \text{speed}$$
$$f = \text{frequency}$$
$$\lambda = \text{wavelength}$$

Phase: this identifies the point in a wave's cycle. A complete oscillation of the sine wave shown in Fig 2.8 is 360°. The phase at point X is then 0°, at Y it is 90° and at Z it is 270°.

Table 2.1

Sub-multiple	Prefix	Symbol
10^{-3}	milli	m
10^{-6}	micro	μ
10^{-9}	nano	n
10^{-12}	pico	p

Multiple	Prefix	Symbol
10^{3}	kilo	k
10^{6}	mega	M
10^{9}	giga	G

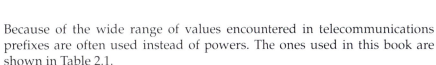

Fig 2.8 Phase and the wave.

Because of the wide range of values encountered in telecommunications prefixes are often used instead of powers. The ones used in this book are shown in Table 2.1.

The electromagnetic spectrum

The word radio comes from the latin word 'radius' meaning a spoke or ray.

This is a most important group of waves whose existence was originally postulated by Maxwell in 1867. He predicted the speed of these waves and the experimental value obtained for light was almost identical to their calculated value. This strongly suggested that light itself was an electromagnetic wave. Other types of wave were found to show similar properties to the radio waves produced by Hertz. The names originally given to the different types referred to their method of production (e.g. gamma rays were emitted from radioactive nuclei). The complete range of observed frequencies of electromagnetic waves is called the **electromagnetic spectrum**. The sections of interest for telecommunication purposes are those of radio and infrared.

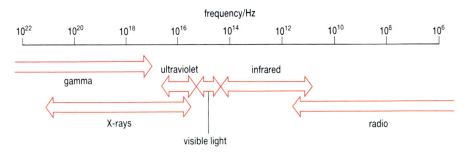

Fig 2.9 The electromagnetic spectrum.

It will be noticed that some of the regions overlap. For instance the lowest frequency infrared waves overlap with the highest frequency radio waves. Within this overlap region the waves are exactly the same – the only difference is the method of their production. The same detector could be used for each of them.

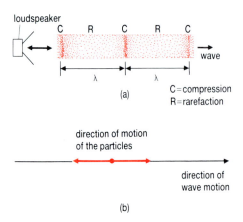

Fig 2.10 (a) A sound wave. (b) Longitudinal wave motion.

Sound waves

Sound waves are different to the waves on the surface of water – they are longitudinal waves. With **longitudinal** waves, the vibrations of the particles in the medium are parallel to the direction in which the wave travels. This produces compressions (regions of higher pressure) and rarefactions (regions of lower pressure) that correspond to the crests and troughs of the transverse wave. Sound waves can only travel a very limited distance and are of little use for telecommunication purposes. Even changing them into electromagnetic waves of the same frequency is not practical. In an analogue telecommunication system they have to be encoded onto electromagnetic waves of much higher frequency. This process is discussed in Section 2.4. The sound frequencies that can be detected by the human ear are often taken as within the range 20 Hz to 20 kHz, though this range varies among individuals and the upper limit decreases with age.

Bandwidth

The word bandwidth is used in two ways. One refers to the actual range of frequencies that is to be transmitted and this is called the **signal bandwidth**. It also refers to the complete range of frequencies that can be transmitted through a telecommunication channel. This is called the **channel bandwidth**.

To transmit the complete range of frequencies that can be detected by the human ear would require 20 kHz of signal bandwidth. If the channel bandwidth is 100 kHz then it would be possible to send only five conversations (the ways in which this can be done are described in Section 2.5). With the telephone system, only the frequencies from 300 to 3400 Hz (i.e. the signal bandwidth is 3.1 kHz) are actually transmitted. However, it is fairly easy to recognise the voice of the person at the other end and the messages sent are completely intelligible. The reason is that most of the frequencies used in speech lie within this range. In practice a bandwidth of 4 kHz is allocated per voice channel in the telephone network. The extra 900 Hz is used to keep the channels separate. With the original 100 kHz bandwidth it is then possible to send twenty-five voice conversations simultaneously along the same transmission path, each channel being allocated its own unique 4 kHz bandwidth.

PRINCIPLES OF TELECOMMUNICATIONS

Practical waveforms

In practice waveforms are very rarely simple sinusoidal waves. If two musical instruments are played at the same pitch it is usually quite easy to distinguish between them. The reason is that, in addition to the fundamental frequency that determines the pitch, there are **overtones**, which are multiples of the sinusoidal fundamental frequency, present. The actual overtones and their amplitudes will differ for each instrument and determine the quality (or timbre) of the note played. If a note of pitch of 1024 Hz is played then it may contain overtones of multiples of 1024 Hz e.g. 2048 Hz, 3072 Hz, 4096 Hz and so on. Not all possible overtones may occur and the higher the frequency of the overtone the smaller its amplitude. The resultant waveform will be the sum of the individual sine waves. If the maximum frequency that can be transmitted along a communication channel is 3400 Hz then frequencies higher than this will not be transmitted. The received musical sound will then sound somewhat distorted. With speech transmission overtones are also produced but, as the fundamental and majority of overtone frequencies involved are lower than 3.4 kHz, less distortion is caused.

INVESTIGATION

Synthesis of waveforms from sine waves

By adding sine waves it is possible to produce more complex waveforms. This investigation suggests a number of ways of doing this.

You will need: an oscilloscope;
two signal generators;
a loudspeaker.

1. Set up the apparatus as shown in Fig 2.11. It may be more satisfactory to keep the loudspeaker disconnected except when listening for the difference in sounds.

2. Using the oscilloscope set one signal generator to 100 Hz (the fundamental frequency) and the second to 200 Hz (the overtone) at about 20 percent of the amplitude of the 100 Hz signal.

3. With both signal generators switched in observe the resultant waveform. Due to phase differences between the two signal generators the resultant trace on the oscilloscope may not be stationary. With the loudspeaker switched in listen to the difference between the fundamental frequency on its own and the fundamental plus overtone.

4. Investigate the effects of varying separately **(a)** the amplitude and **(b)** the frequency of the overtone.

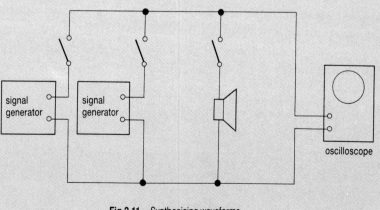

Fig 2.11 Synthesising waveforms.

Square waves

It can be shown mathematically that any repetitive waveform of frequency f can be considered to be made up of a number of sine waves having frequencies f, $2f$, $3f$, $4f$ and so on. As with musical notes, the fundamental frequency is f and the overtones are $2f$, $3f$, $4f$, etc. Square waves, which are of great importance in telecommunications for the transmission of digital signals, are examples of repetitive waveforms. A **square wave** of frequency f can be shown to be made up from sine waves of frequencies f, $3f$, $5f$ and so on. As the frequency of the overtone increases so its amplitude becomes smaller. In Fig 2.12 only the fundamental and the first two overtones are

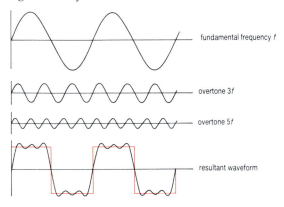

fundamental frequency f

overtone $3f$

overtone $5f$

resultant waveform

Fig 2.12 Producing a square wave.

shown. If these are added together then the resultant wave is fairly close to the shape of the original square wave. If square waves are transmitted via a normal analogue voice channel then only frequencies below 3.4 kHz will be transmitted. In practice this means that only low frequency digital pulses can be transmitted along analogue links.

Bandwidth requirements

With music the range of frequencies that need to be transmitted is much larger than that allowed for in a voice channel, as the frequencies produced by musical instruments cover a much wider frequency range than normal speech. A bandwidth of 15 kHz is required for good quality music reception. To transmit the signals for a television transmission a bandwidth of 6 MHz is required.

QUESTIONS

2.5 Attempt the following question on electromagnetic waves if you are not familiar with the prefixes used or the use of powers. Take the speed of electromagnetic waves as $3 \times 10^8 \, \text{m s}^{-1}$.
(a) If the wavelength is 3 m, calculate the frequency.

2.4 MODULATION

In telecommunication systems the information signals are not transmitted at their actual frequencies. If they were, only one signal could be transmitted along the channel at a time. The signals are encoded onto a much higher frequency signal called a **carrier**. The carrier actually transfers the information along the transmission path. The carrier is changed according to the value of the information signal. This process is called **modulation** and a number of different types exist. At the receiving end **demodulation** occurs and the signal is extracted from the carrier.

Amplitude and frequency modulation

These methods of modulation are used in radio and TV transmissions. The carrier wave is produced by the electronic circuitry and the information to be transmitted is superimposed in one of two ways. Amplitude modulation varies the amplitude of the carrier to carry the information. Frequency modulation varies the frequency of the carrier to do the same thing. These two methods are discussed further in Section 3.3.

Pulse code modulation

The principles behind PCM were established in 1937 by Alec Reeves. However PCM could not be implemented in a practical system until the development of semiconductor circuitry had sufficiently advanced. This occurred in the early 1960s.

This method involves 'sampling' the amplitude of the analogue signal at regular time intervals using electronic switches. This is called **pulse amplitude modulation** (PAM). These analogue samples are then turned by an ADC into binary numbers which are then transmitted. This latter process is called **pulse code modulation** (PCM). In one method of PCM the amplitude is assigned one of 128 levels (decimal 127 corresponds to the

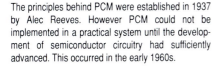
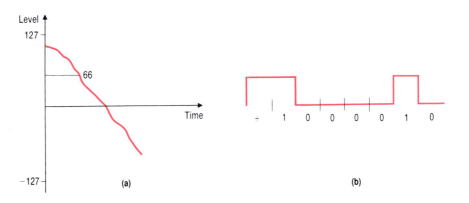

Fig 2.13 Pulse code modulation **(a)** sampling the signal **(b)** the encoded signal.

seven digit binary number 1111111) and the eighth bit depends on whether it is positive or negative. In Fig 2.13(a) the level at the point on the signal being measured is 66 (binary 1000010). If the value is positive it is assigned a '1' or, if negative, a '0'. This eight bit number is then transmitted. This could be sent as a series of pulses as illustrated in Fig 2.13(b). At the receiving end the reverse of the encoding procedure occurs using a DAC.

With PCM the sampling frequency has to be high enough to ensure that the original signal can be satisfactorily decoded at the other end. For this to occur the sampling frequency must be at least twice the maximum frequency present in the transmitted signal. With a telephone voice circuit the signal bandwidth is 4 kHz and so the sampling frequency is taken as 8 kHz. The number of binary digits transmitted per second for a single channel is called the **bit rate**.

To transmit high quality music using PCM the bit rate must be much larger. For good quality reception 16 bits are required to encode each sample and a bandwidth of 15 kHz must be transmitted. The bit rate is given by:

Bit rate = $16 \times 30\,000 = 4.8 \times 10^5$ bit s^{-1}.

Television requires a bit rate of 70 Mbit s^{-1}.

$$\text{Bit rate} = \text{number of bits per sample} \times \text{sampling frequency}$$
$$= 8 \times 8000$$
$$= 64\,000 \text{ bit s}^{-1} \text{ or } 64 \text{ kbit s}^{-1}$$

One of the simplest forms of amplitude modulation is used in digital optical fibre transmissions which use light or infrared waves as the carrier and PCM. A '1' is represented by light being emitted by the laser source (maximum amplitude) and a '0' by an absence of light (zero amplitude). This is illustrated in Fig 2.14.

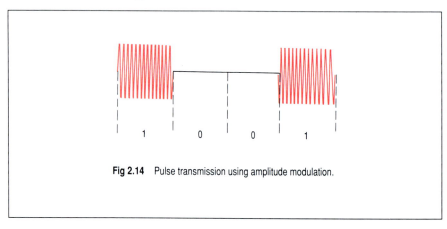

Fig 2.14   Pulse transmission using amplitude modulation.

Phase shift keying

There are a number of variations used in this type of modulation and one of them is shown in Fig 2.15. With **phase shift keying** (PSK) the digital pulses produced by PCM are used to change the phase of the carrier wave. If a '1' is being transmitted there is no phase change but a '0' produces a phase change of 180°. In Fig 2.15 the binary number 1001 is being transmitted. At the receiving end the signal is demodulated to retrieve the original binary digits.

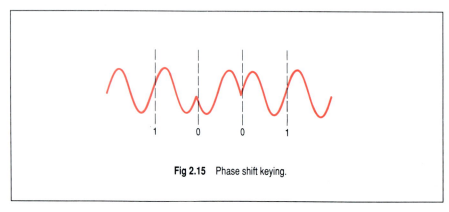

Fig 2.15   Phase shift keying.

PRINCIPLES OF TELECOMMUNICATIONS

QUESTIONS

2.7 A proposed PCM system is designed to work with a total of 64 levels and to transmit a 4 kHz voice channel.
  (a) How many bits are required to transmit each sample? (Hint: express 64 as a power of 2.)
  (b) What is the bit rate?
  (c) If an extra bit for checking purposes has to be added to each sample sent what would be the total bit rate?
  (d) What advantage do you think there would be of using fewer levels than the standard 256 levels?
  (e) How do you think the accuracy of the decoded signal would compare between a 64 level and a 256 level system?

2.8 Consider a signal generated at a frequency of 1 GHz.
  (a) What time period does one cycle occupy?
  (b) With phase shift keying, how long would it take to send one bit?
  (c) How long would it take to send the 64 kbits from a 4 kHz bandwidth signal?

2.5 MULTIPLEXING

The very early telecommunication systems allowed only one message to be sent at a time. Efficiency could obviously be increased by sending a number of messages simultaneously along the same channel. The process is called **multiplexing**. A number of ways were devised of doing this but only with the coming of modern electronic techniques did it become possible to send large numbers of messages simultaneously along a transmission channel. A coaxial cable can have a bandwidth of 60 MHz. Since the normal voice channel requires 4 kHz, theoretically a very large number can be sent along the same coaxial cable. In practice the maximum number that can be transmitted simultaneously along such a cable is 10 800.

Frequency division multiplexing

This example is only meant to show the features of **frequency division multiplexing** (FDM). Assume that there are three telephone signals to be sent simultaneously along a channel. Each one is allowed 4 kHz of bandwidth. If all were sent at their normal frequencies then the received signals would be completely confused. If signal 1 is transmitted at its normal frequencies, signal 2 in the range 4 to 8 kHz and signal 3 in the range 8 to 12 kHz, then all three signals can be sent at the same time with a

In 1936 the first FDM telephone cable came into operation between Bristol and Plymouth. It carried 12 channels.

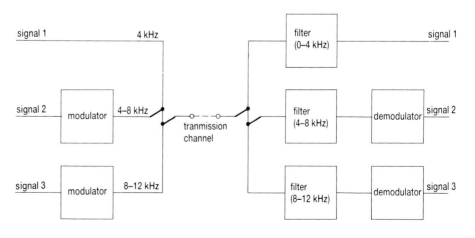

Fig 2.16 Frequency division multiplexing.

channel bandwidth of 12 kHz. This is illustrated in Fig 2.16. A modulator is required for those signals not transmitted at their natural frequencies i.e. signals 2 and 3. At the receiving end there are filters that respond to certain ranges of frequencies. For example, the filter for signal 2 will only allow the frequencies from 4 to 8 kHz through. The filters separate the three signals which are then passed through demodulators (not required for signal 1) to reproduce the original signals.

In practice 12 voice channels are formed into a 'group' and the transmitting frequencies used are much higher than those illustrated but the principles are the same. 12 voice channels will occupy 48 kHz of bandwidth. For long distance telephone transmissions, five 12-channel groups will be joined together to form one supergroup. Supergroups can be further combined until the total is 10 800 channels. FDM requires analogue signals and is used extensively at present in the telephone network. As digital transmissions become more prevalent this method will be superseded by time division multiplexing.

Time division multiplexing

As long ago as 1874 the first method of TDM was invented by Edison. His device allowed four messages to be sent simultaneously, two in each direction, though it never became a practical system.

In the normal telephone signal encoded into PCM there are 8000 samples sent every second or one sample every 125 μs (1/8000). But if a pulse (either high or low) is produced every 1 μs then the time needed to send

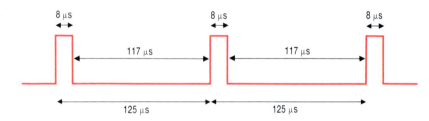

Fig 2.17 Time division multiplexing.

one sample of eight bits is 8 μs. This leaves an interval of 117 μs between each sample as shown in Fig 2.17. With the appropriate electronics it is possible to insert extra channels into these spaces. This process is called **time division multiplexing** (TDM). With the example shown, it is theoretically possible to send 15 separate channels (125/8 = 15.6). In practice extra bits have to be sent in addition to the eight in the sample to enable each channel to be accurately decoded at the other end and to detect any errors that occur in transmission. This reduces the number of separate channels that can be sent.

QUESTIONS

2.9 This question involves the same 64 level PCM system discussed in question 2.7.
 (a) If the pulse length is 1 μs what is the total time required to send each sample (including the checking bit)?
 (b) What is the time interval between each sample?
 (c) How many voice channels could theoretically be sent simultaneously?
 (d) Suggest a method whereby the number of voice channels could be increased.

2.10 Assuming a 4 kHz voice channel and that each sample requires 10 bits (pulse width is 0.1 μs) for encoding, calculate the theoretical number of channels that could be sent using TDM.

2.6 ATTENUATION AND NOISE

Noise

Attenuation of a signal occurs when it is propagated along a communications channel as energy is lost in various ways to the surroundings. Consequently the signal power gets smaller and smaller. At the same time noise is added to the signal. Noise is a problem in all telecommunication systems. Noise has a number of sources although all of these are electrical in nature. It can be classified as natural or artificial. **Natural noise** includes cosmic noise (from outer space) and atmospheric noise (from storms and other electrical effects in the upper atmosphere). The noise produced in electronic circuits (mainly due to the random motion of the atoms and electrons) is also classed as natural. As the noise generated in electronic circuits increases with temperature, modern high gain amplifiers such as those used in satellite telecommunication links are run at low temperature. **Artificial noise** is mainly produced by electrical devices such as motors and switches.

As the signal travels from the transmitter to the receiver, noise energy is being added to an attenuating signal. If nothing is done to boost the transmitted signal then eventually it would disappear amidst the noise. The purpose of the repeaters is to increase the power of the signal and, of the regenerators, to produce new pulses. With repeaters the noise already added to the signal is amplified whereas with regenerators a completely new pulse is generated that is noise-free. At all times the signal must be a minimum number of times larger than the noise, if it is to be satisfactorily decoded at the receiver. Before considering this aspect it is necessary to discuss how signal power is calculated.

The decibel

Power levels in telecommunication systems are expressed in units called decibels. Although this method may appear to be difficult to understand at first, you will find that it makes calculations very easy.

The output power level P_2 is related to the input power level P_1 by means of a logarithmic ratio. This ratio is called the bel B after the inventor of the telephone. In practice the bel turned out to be rather a large unit and it was replaced by 1/10 of a bel or **decibel** dB.

$$\text{Number of dB} = 10 \log_{10}(P_2/P_1)$$

Application to a system

If signals are sent through a system there are a series of gains and losses. A gain (or amplification) is represented by a positive dB value and a loss (or attenuation) by a negative one. This is illustrated in the following examples.

Example 1

The input to an amplifier is 10 mW and the output is 1 W. Calculate the gain.

$$\text{Number of dB} = 10 \log_{10}(1/10^{-2}) = 10 \log_{10}(10^2) = 20\,\text{dB}$$

This is a gain as the ratio is positive.

Example 2

The input power of a signal to a transmission path is 100 mW and the output at the other end is 1 mW. Calculate the loss.

$$\text{Number of dB} = 10 \log_{10}(10^{-3}/10^{-1}) = 10 \log_{10}(10^{-2}) = -20\,\text{dB}$$

This is a loss as the ratio is negative. The loss in a transmission path is sometimes expressed in dB km^{-1} as in the following example. When a series of gains and losses takes place, the resultant power level (in dB) is equal to their algebraic sum.

Example 3
Calculate the dB output of the system shown in Fig 2.18 and calculate the actual output power.

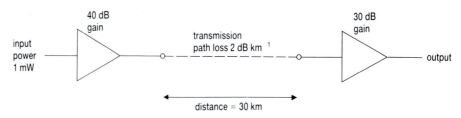

Fig 2.18   Gains and losses in a system.

Transmission path loss $= 2 \times 30 = 60$ dB .

This is negative as it is a loss.

$$\text{dB output} = +40 - 60 + 30 = +10 \text{ dB}$$

The actual power output P is given by;

$$10 = 10 \log_{10}(P/10^{-3})$$

Hence $P = 10^{-2}$ W

Signal-to-noise ratio

The problem of noise imposes limits on how far the signals can be transmitted before it is necessary to pass them through a repeater or regenerator. The transmitted signal power must always exceed the noise power by a certain amount. This is called the **signal-to-noise** (S/N) ratio and is expressed in decibels. If the signal power falls below this value then the received signal cannot be accurately decoded. For example, the S/N ratio might be 20 dB. This means that the transmitted signal must always be larger than the noise by a factor of 100. This is very important in long distance transmission paths which have repeaters (or regenerators) in them. They must be placed so as to ensure that the attenuating signal does not fall below the minimum S/N ratio.

In practice the capacity of a channel using PCM/TDM techniques is limited. Claude Shannon in 1948 developed a relationship between capacity C, bandwidth B and the signal-to-noise ratio S/N. The theoretical maximum for C is given by:
$C = B \log_2 (1 + \text{S/N})$
This means that the capacity of a 4 kHz channel with S/N of 20 dB (S/N = 100) would be 26.7 kbit s^{-1}. In practice it turns out to be a lot less than this.

QUESTIONS

2.11 With telecommunication systems there is a reference power level with which the actual signal (in dB) can be compared. For example, this is 1 mW for coaxial and optical fibre systems and 1 W for satellite systems.
  (a) What is the power carried by a signal 20 dB above the reference level in an optical fibre system?
  (b) In a satellite system, the power of a transmitted signal is 10 dB below the reference level. Calculate the actual power.

2.12 In a particular digital telecommunication system the minimum signal-to-noise ratio is 21 dB.
  (a) If the noise power is 10^{-20} W, calculate the minimum power that the transmitted signal can have.
  (b) If the initial signal power is 10^{-3} W, what is the maximum distance the signal can be propagated along a transmission path of attenuation rate 2.0 dB km^{-1} before it is necessary to amplify it?
  (c) What would happen to the signal if the minimum signal-to-noise ratio were exceeded?

SUMMARY

The different types of telecommunication system use similar sub-systems. Both analogue and digital signals are used at present though all systems are changing to digital. The following types of modulation are important in telecommunication systems: amplitude, frequency and pulse code modulation together with phase shift keying. There are two main methods of multiplexing the signals: frequency division multiplexing (each signal is allocated a different frequency range and the modulated signals are then transmitted as analogue signals) and time division multiplexing (the samples from each signal are slotted into vacant time intervals on the carrier as digital signals). The transfer of signals involves attenuation and the addition of noise to the signal. Amplification (for analogue signals) or regeneration (for digital signals) must be provided at various stages to ensure that the signal strength does not fall below the minimum signal-to-noise ratio.

QUESTIONS

2.13 (a) Why is the process of modulation so important in telecommunication systems?
(b) Explain why multiplexing is essential for transmitting large numbers of signals simultaneously along a single channel.
(c) Describe the similarities and differences between frequency division and time division multiplexing.

2.14 Transmitting a television signal requires a bandwidth of 6 MHz.
(a) If this is to be encoded using pulse code modulation, how many samples per second would be needed?
(b) What would be the number of bits required, assuming each sample required 8 bits?
(c) If using phase shift keying what bandwidth would be required on the carrier?

2.15 Fig 2.19 shows an analogue signal that is to be encoded using pulse code modulation. The encoded signal has seven levels each being represented by a three bit number.
(a) What is the decimal value of the level each time the signal is sampled? Change these into their binary equivalents.
(b) Assume the pulses are transmitted without any changes. Using a similar grid, plot out the received signal. Explain the difference between the original signal and the decoded version. How could you minimise the difference between them?

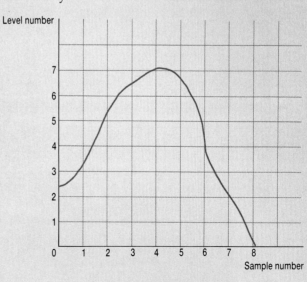

Fig 2.19

2.16 In a simple 4-bit digital communication system, binary numbers are transmitted as logical levels (0 or 1) for periods of 1 μs following immediately one after another, the most significant bit coming first.

(a) Draw a graph of the signal representing the binary number 1011.

(b) An audio signal is digitised in this system with a sampling interval of 100 μs giving the following sequence of binary values repeated over and over again.

1010 1101 1110 1011 0110 0011 0010 0101

Mark the corresponding analogue values on a graph against time, and draw a reasonable form for the audio signal.

(c) What is the frequency of the audio signal?

(d) Approximately what bit rate would be required for the transmission of the digital signal in this form? Explain how you arrived at your answer.

(e) State two ways in which the system might be changed so as to transmit more information about the audio signal.
(Adapted from UCLES spec.)

2.17 A 1 mW signal is sent into the telecommunication system shown in Fig 2.20.

(a) Calculate the overall gain of the system.

(b) Calculate the output power.

(c) If the background noise power is 10^{-12} W and the signal-to-noise ratio must exceed 20 dB, can the signal be satisfactorily decoded at the other end?

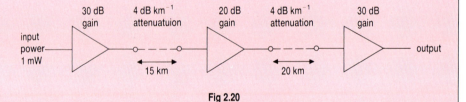

Fig 2.20

Theme 2

RADIO COMMUNICATIONS

The use of radio waves for telecommunication purposes dates back a hundred years. During this time there has developed a sophisticated telecommunication system that covers the entire planet together with a source of entertainment and information that is available almost everywhere. This theme is divided into two chapters with the first looking at the transmission of radio waves and the second at their reception. It starts by looking at the basic properties of radio waves and the principles behind communication by radio. This is followed by the modulation of the signals and the propagation of radio waves. The second chapter looks at the structure of the most basic radio receiver and demodulation of the transmitted signals. Straightforward improvements that can be made to improve the reception are then considered. The use of aerials in transmitters and receivers is discussed. Most of the theme is linked to radio broadcasting and reception but the final section looks at other uses of radio waves in telecommunications.

PREREQUISITES

Before you study this section you should have some familiarity with the following:
- The nature and properties of electromagnetic waves.
- The action of inductors and capacitors in a.c. circuits.

The photographs show a modern combined transmitter and receiver (or transceiver) and a transmitter made in 1896.

Chapter 3

TRANSMITTING RADIO WAVES

LEARNING OBJECTIVES

After studying this chapter, you should be able to:

1. know that radio waves are electromagnetic waves and show the usual wave properties including polarisation;

2. recall the basic building blocks of a simple radio transmitter and receiver;

3. describe how amplitude modulation and frequency modulation are used in radio transmissions;

4. know that, for sinusoidal waves, the bandwidth of the carrier wave is twice the maximum modulating frequency;

5. describe the part that the ionosphere plays in the propagation of radio waves;

6. know that transmitted radio waves can reach the receiver as ground, sky or space waves.

3.1 RADIO WAVES

Fig 3.1 Heinrich Hertz who discovered radio waves.

It has been mentioned that the first direct evidence of the existence of electromagnetic waves was the result of Hertz's experiments in 1888. In this chapter and the next we will be looking at how the radio waves discovered by Hertz are used for telecommunications.

The nature of electromagnetic waves

If electrical charges are accelerated or decelerated then electromagnetic waves are produced. Perhaps the simplest example is that of a metal wire connected to an a.c. (alternating current) supply. The oscillations of the electrons in the conductor will produce **electromagnetic waves** of a frequency equal to that of the a.c. supply. These waves are transverse and travel away from the conductor at a speed that is the same as that of light. Maxwell showed that the speed of electromagnetic waves is given by the formula:

$$c = 1/\sqrt{\mu\varepsilon} \quad \text{where } \mu = \text{permeability of the medium}$$
$$\varepsilon = \text{permittivity of the medium}$$

The permeability and permittivity are constants associated with magnetic and electrical fields respectively. More details on these can be found in a basic physics book such as the core book *Physics* Chapters 16 and 18. Substituting the appropriate values in free space (i.e. a vacuum) leads to the following value of c: 3×10^8 m s^{-1}. In any other medium the appropriate values for that medium will need to be substituted.

Although transverse, electromagnetic waves differ from water waves in a number of ways. Electromagnetic waves actually consist of two waves: an

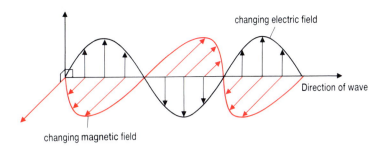

changing electric field

Direction of wave

changing magnetic field

Fig 3.2   The electromagnetic wave.

electrical wave (often called the E-field) and a magnetic wave (often called the B-field) that travel in phase but at right angles to each other. This is illustrated in Fig 3.2. Another difference is that electromagnetic waves do not require a medium for their transmission. In fact, electromagnetic waves travel fastest in a complete vacuum.

Properties of waves

All waves show, under suitable conditions, a number of basic properties. These are **reflection**, **refraction**, **diffraction** and **interference**. Fig 3.3 illustrates each of these using plane waves. These properties are important in understanding the ways in which radio waves are propagated.

In addition to these properties, transverse waves show an additional property – **polarisation**. This means that the wave is restricted to travelling in one plane. Although electromagnetic waves always consist of an electrical wave and a magnetic wave in phase with each other, it is usual to consider just the electrical component when referring to polarisation. If the electromagnetic waves produced by the oscillations of electrons in a vertical wire (such as a transmitting aerial) are considered, then the electrical wave travels in a vertical plane. The electromagnetic wave is said to be polarised in a vertical plane. The receiving aerial must also be positioned vertically to receive the waves as it only detects the electrical component of

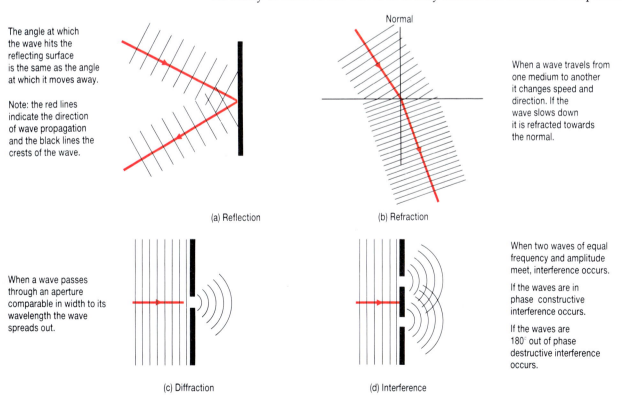

The angle at which the wave hits the reflecting surface is the same as the angle at which it moves away.

Note: the red lines indicate the direction of wave propagation and the black lines the crests of the wave.

(a) Reflection

When a wave travels from one medium to another it changes speed and direction. If the wave slows down it is refracted towards the normal.

Normal

(b) Refraction

When a wave passes through an aperture comparable in width to its wavelength the wave spreads out.

(c) Diffraction

When two waves of equal frequency and amplitude meet, interference occurs.

If the waves are in phase constructive interference occurs.

If the waves are 180° out of phase destructive interference occurs.

(d) Interference

Fig 3.3   The basic properties of waves.

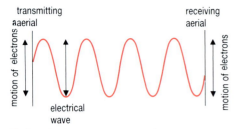

transmitting aerial

receiving aerial

motion of electrons

motion of electrons

electrical wave

Fig 3.4 A polarised electromagnetic wave.

the wave. The electrons in the receiving aerial oscillate at the same frequency as the incoming electrical wave. If the electromagnetic waves were emitted in all possible directions to the vertical then the electromagnetic waves would be unpolarised. The receiving aerial could only pick up those waves with an electrical component parallel to the direction of the aerial.

Radio section of the electromagnetic spectrum

Although **radio waves** form only a section of the electromagnetic spectrum they are very important in telecommunications. The range of frequencies over which radio waves can be produced is quite large as can be seen from Table 3.1 and it has been sub-divided into a number of sections called **frequency bands**. Radio waves with a frequency above 1 GHz are usually referred to as **microwaves**. It is possible to use either frequency or wavelength when discussing waves as they are related by the formula: speed = frequency × wavelength. It has become standard to use frequencies when discussing radio waves and this practice will be used.

Why is the radio section of the electromagnetic spectrum of so much importance for communication purposes? The main reasons are that radio waves can be produced at accurately maintained frequencies and that they can travel long distances. (Modern technology now allows accurately maintained frequencies to be produced in the infrared region using lasers. This will be discussed in Section 6.1)

Table 3.1   The radio spectrum

Classification	Abbreviation	Frequency range	Wavelength range
very low frequencies	VLF	3–30 kHz	100–10 km
low frequencies	LF	30–300 kHz	10–1 km
medium frequencies	MF	300–3000 kHz	1000–100 m
high frequencies	HF	3–30 MHz	100–10 m
very high frequencies	VHF	30–300 MHz	10–1 m
ultra high frequencies	UHF	300–3000 MHz	100–10 cm
super high frequencies	SHF	3–30 GHz	10–1 cm
extra high frequencies	EHF	30–300 GHz	1–0.1 cm

QUESTIONS

3.1 (a) Look at a fixed TV aerial. From the orientation of the aerial is it receiving polarised signals? If so, in which plane is the wave polarised?

(b) Why is it so important that the frequencies used in radio transmissions should be accurately maintained?

3.2 (a) Radio broadcasting in the VHF band is restricted to the range 88 to 108 MHz. Calculate the maximum and minimum wavelengths used.

(b) If 10 kHz bandwidth were allowed for each channel in radio communication how many channels would be available in the complete (i) MF band and (ii) VHF band?

(c) Using the *Radio Times*, find out in which frequency bands Radio 1 and Radio 4 are broadcast.

(d) By looking at the frequency ranges on your radio, estimate what fraction of the (i) LF, (ii) MF and (iii) VHF bands can actually be received.

Early transmissions

As mentioned in Section 1.3 the first transmissions used an on/off method in conjunction with the Morse code. This type of radio transmission is sometimes called 'carrier keying' as the carrier wave is continually switched on and off. Through the action of the induction coil the consequent spark discharge (hence their name – spark transmitters) produced a range of high frequency currents that caused electromagnetic waves to travel away from the aerial. The receiver picked up these signals and, with the aid of the 'coherer' (the action of this device is studied in question 4.1), the transmitted information could be decoded. As there was no physical link between the transmitter and receiver this allowed mobility of either or both stations. However the efficiency of the radio system was limited for a number of reasons:

- The rate of information transfer was slow as the message had to be translated into Morse code. Means were needed of directly encoding speech onto radio waves. As with the telephone this would allow untrained people to communicate directly with each other.

- To increase the distance between transmitter and receiver a way of amplifying the signal before and after transmission was required.

- The radio waves produced in early transmissions contained a wide range of frequencies. To allow a number of transmissions to take place in the same area it was necessary to broadcast radio waves at fixed frequencies.

INVESTIGATION

A simple transmitter

You will need: a battery and holder;
SPST (single pole single throw) switch;
portable radio receiver.

In this investigation you will use a primitive form of transmitter together with a conventional radio receiver. Opening and closing the switch produces a wide frequency range of radio waves. The aim is to test the operating characteristics of the 'transmitter'. Normally a licence is required for any type of radio transmission but, as the range of transmission is very limited due to the low voltages used with the transmitter, this is not necessary.

Fig 3.5   (a) A simple transmitter and radio receiver.

Fig 3.5   (b) The first radio broadcasting transmitter in the UK.

1. Set up the apparatus as shown in Fig 3.5 (a). Set the receiver to the MF band and tune it so that no station is being received. Open and close the switch. The radio should produce a noise at the moment the switch is opened or closed. It may be necessary to rotate the receiver to obtain satisfactory results.

Improvements in transmission and reception

Although the early experimenters devised ingenious ways of overcoming difficulties in transmission and reception it was the invention of the diode valve and then the triode valve that produced the greatest developments. The diode allowed demodulation (often called 'detection') of the signal (more details are given about this in Section 4.2). Also it could be used for rectification of mains a.c. as valves used high d.c. voltages. The triode enabled signals to be amplified both in the transmitter and in the receiver. Using suitable circuits triodes could produce fixed frequency signals. Circuits which do this are called **oscillators**. This allowed a number of simultaneous signals at different frequencies to be transmitted and received in the same area. Developments allowed higher frequencies to be produced and it was found that these signals travelled further.

Up until the end of World War 1 radio was only used for communication purposes, e.g. ship to shore and military communications. Radio for entertainment purposes did not really start until 1920. Whereas in the USA a large number of independent broadcasting companies came into existence, in the UK there was only one. This was the BBC and it became responsible for all radio broadcasting.

When fixed frequency broadcasts became possible the frequencies above 1 MHz were thought to be 'useless'. They were allocated for the use of amateur radio enthusiasts who found that these waves could be used for long distance communication – even to the other side of the world.

The basic radio system

The radio system has two sections – the transmitter and receiver.

The radio transmitter

- The microphone changes the sound waves into electrical currents. The frequency of these a.c. currents corresponds to the frequencies of the original sound waves. These are usually called **audio frequency** (a.f.) signals.

- The oscillator is an electronic device that produces an a.c. signal at the required frequency for transmission. This is the **radio frequency** (r.f.) signal and is at a much higher frequency than the a.f. signals. This becomes the carrier wave for the signals.

- The modulator encodes the a.f. signals onto this fixed frequency signal. The process is explained further in Section 3.3. The modulated signal is amplified in the r.f. amplifier before producing alternating currents in the aerial. These currents then produce the electromagnetic waves that travel away from the transmitting aerial.

The radio receiver (Receivers will be considered further in Section 4.2).

- The receiving aerial picks up the transmitted radio waves (plus many others). These induce alternating currents in the aerial corresponding to the transmitted frequencies.

- The tuning circuit selects the desired r.f. signal.

- In many types of radio receiver the r.f. signal is amplified before passing to the next stage. In very simple receivers there is no r.f. amplifier.

- The demodulator removes the r.f. component of the signal leaving the original a.f. component.

- The amplifier increases the size of the a.f. signal before feeding it to the loudspeaker where these a.f. signals are changed back into sound waves that correspond to the original ones entering the microphone.

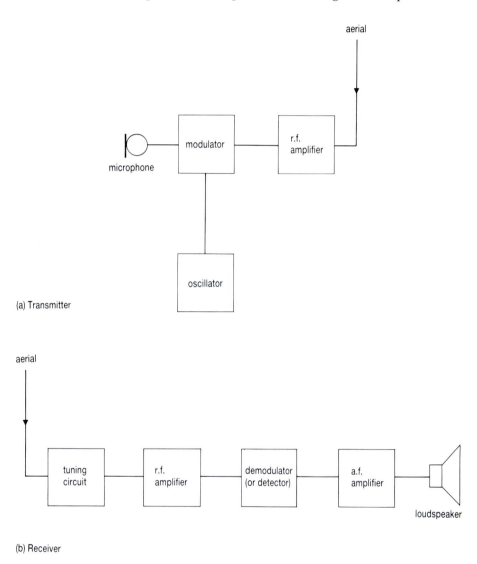

(a) Transmitter

(b) Receiver

Fig 3.6 The basic radio system.

As already mentioned the a.f. waves have to be encoded onto an r.f. carrier. The process of encoding is called modulation. The two main ways of modulating radio waves will be described in this chapter – amplitude modulation and frequency modulation.

Amplitude modulation (AM)

Amplitude modulation encodes the a.f. waves onto an r.f. carrier wave by varying its amplitude. Figure 3.7 shows how the a.f. signal changes (modulates) the amplitude of the carrier wave. The amplitude 'follows' the changes in the modulating signal.

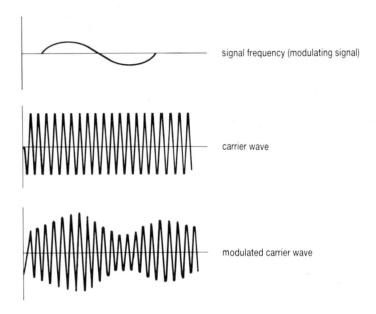

signal frequency (modulating signal)

carrier wave

modulated carrier wave

Fig 3.7   Amplitude modulation.

Transmitting AM radio waves

The unmodulated radio carrier wave is a sinusoidal wave produced at a constant frequency and the modulating a.f. wave is used to vary its amplitude. It is possible to show mathematically that the transmitted carrier behaves as though it occupied a range of frequencies. If the carrier frequency is f_c and the maximum frequency of the modulating signal is f_m, then the maximum transmitted frequency is equal to $(f_c + f_m)$ and the minimum transmitted frequency is $(f_c - f_m)$. The transmitted frequency range, or bandwidth, is given by the following expression:

$$\text{Bandwidth} = (f_c + f_m) - (f_c - f_m)$$

This relationship is only true if both the carrier and the modulating signals are sinusoidal. The expression simplifies to:

$$\text{Bandwidth} = 2f_m$$

Although the range of frequencies that can be detected by the human ear can extend from 20 Hz to 20 kHz, there are bandwidth problems associated with transmitting this range of frequencies just as with the voice channels in the telephone system. To transmit as many AM radio broadcasts as possible in the allowed broadcasting bands a restricted range of frequencies is transmitted. In the UK the range of frequencies transmitted in each channel in the LF and MF bands is from 50 Hz to 4.5 kHz. This means the

R.A. Fessenden managed to broadcast speech in 1906. He used a high frequency alternator to generate continuous radio waves at a frequency of 45 kHz and these were amplitude modulated by the audio frequencies produced through a carbon microphone.

With AM, identical information is carried in each of the sidebands. It is possible to suppress one of the sidebands together with the carrier wave and transmit just the other sideband. This is called single sideband (SSB) transmission. It does require complex and expensive equipment. One place where it is used is with FDM in coaxial cable transmission.

channel bandwidth is 9 kHz. The frequency spectrum of such an AM signal is shown in Fig 3.8.

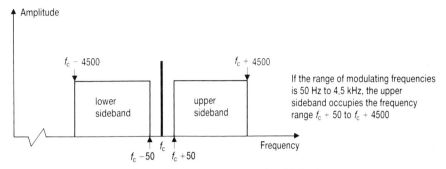

If the range of modulating frequencies is 50 Hz to 4.5 kHz, the upper sideband occupies the frequency range $f_c + 50$ to $f_c + 4500$

Fig 3.8 The frequency spectrum of an AM signal.

Frequency modulation (FM)

With **frequency modulation** it is the frequency of the carrier wave that is changed rather than its amplitude. The larger the instantaneous value of the modulating signal the greater the frequency change from the carrier's original value. For positive displacements of the modulating signal the frequency of the carrier increases and for negative displacements the frequency decreases. This is illustrated in Fig 3.9 with a sinusoidal modulating signal. It can be seen that the amplitude of the carrier does not change. This means that the information encoded on the carrier is not affected by any variations in the carrier's amplitude.

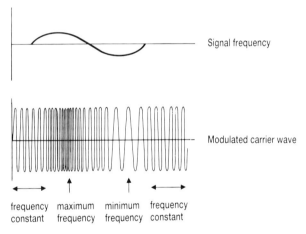

Fig 3.9 Frequency modulation.

INVESTIGATION	**Amplitude modulation and frequency modulation of a carrier wave**

You will need: two signal generators, one of them with AM/FM facilities; oscilloscope.

In this investigation you will use a low frequency a.c. signal to modulate a much higher frequency a.c. signal. Although the frequencies are much lower than those used in r.f. transmissions the principle is exactly the same.

1. For amplitude modulation the apparatus set-up is shown in Fig 3.10(a). One of the signal generators must have an AM input facility and the modulating signal is connected to this AM input. Suggested

values to start with are shown in the diagram. The amplitude of the modulating frequency must be much smaller than that of the carrier frequency.

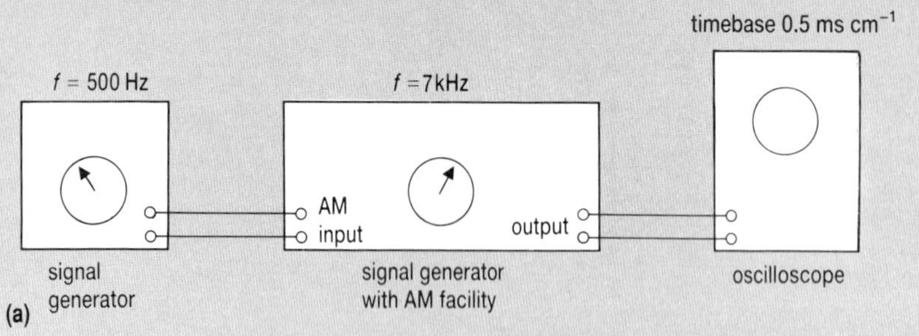

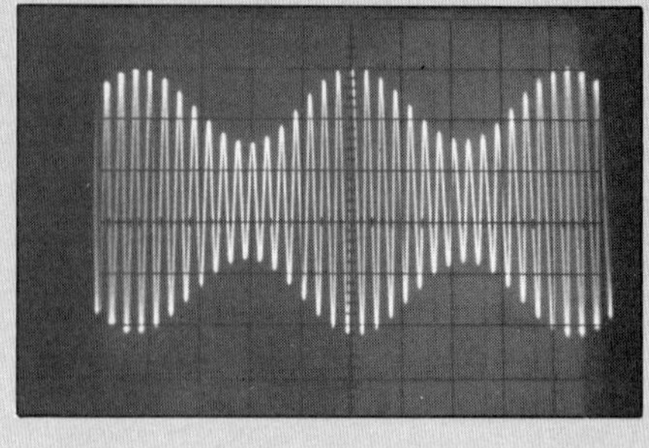

Fig 3.10 **(a)** Demonstrating amplitude modulation.
(b) An AM waveform.

2. Investigate the effects of varying separately (a) the amplitude and (b) the frequency of the modulating signal.

3. If your signal generator produces triangular or square waves you could look at the effects of modulating the carrier signal with these.

4. Connecting a loudspeaker to the output of the second signal generator can produce some interesting effects. You could start with a carrier frequency of 500 Hz and the lowest modulating frequency available from the signal generator. Try to explain what you hear.

5. For frequency modulation the apparatus is used in the same way as for AM but the FM input to the second signal generator is used. Set the modulating frequency to 10 Hz and the carrier frequency to 1 kHz. Investigate the effects of varying the amplitude and frequency of the modulating carrier. Note that the trace observed on the oscilloscope will probably not remain stationary.

Transmitting FM radio waves

When AM radio signals are transmitted on the LF and MF bands, the channel bandwidth used is 9 kHz. Although speech does suffer a little distortion, there is more distortion for musical sounds as the higher frequencies produced will not be transmitted (refer back to Chapter 2 for details). This problem is overcome with FM transmissions where frequencies up to 15 kHz are transmitted. However, not only does this increase the basic bandwidth required but FM produces side frequencies that are multiples of the modulating frequency and these must be transmitted if the received signal is to be satisfactorily decoded.

Calculation of the total bandwidth needed for transmission is complex but its value can be taken as 180 kHz, assuming signal frequencies of up to 15 kHz are being transmitted. The channel bandwidth is 200 kHz. This is why FM broadcasts can only take place in the VHF band where there is sufficient bandwidth to accommodate them.

In the UK the broadcasting frequencies can be spaced 50 kHz apart but adjacent transmitters are separated by a minimum of 200 kHz. The reason for this is that the maximum transmission distance for broadcasts in the VHF band is of the order of 100 km due to the straight line propagation of radio waves at these frequencies (see Section 3.4 for more details). Consequently adjacent transmitters broadcast on very different frequencies while transmitters a long distance apart can use adjacent frequency bands.

Comparing AM and FM transmissions

The limitations of channel bandwidth associated with AM broadcasts has already been mentioned. There is another problem with AM transmissions.

The first VHF broadcasts by the BBC were from Wrotham in Kent in 1955. Although 750 000 receivers were sold within a year, VHF broadcasts were not as popular as had been expected in spite of the much improved reception. The main reason was that the transistor radio had just become available and, although receiving only on the LF and MF bands, its portability made it much more desirable.

While there are many high powered transmitters (e.g. Radio 4 is broadcast in the VHF band from Wrotham at a frequency of 93.5 MHz with a power of 120 kW) there are many low powered transmitters designed for only a very local coverage. These are normally in the VHF band and their power output may be as low as 10 W.

Electrical noise causes changes in the amplitude of the signal. When the received signals are decoded, the audio signals are distorted. Any changes in the amplitude of an FM signal do not cause any changes in the decoded signal as the audio signals are encoded as frequency changes. This means that FM transmissions are less susceptible to the effects of noise and any variations in the received amplitude of the signal. If you listen to the same programme being transmitted on both AM and FM you will notice the difference in quality between them.

However, AM transmissions have two main advantages. The electronic circuitry is cheaper and there are fewer components to go wrong. Also, the transmission range of AM broadcasts is considerably greater than for FM (see Section 3.4). This means that fewer transmitters are required to cover a given area.

Regular stereo broadcasts were started by the BBC in 1966 on what is now called Radio 3. In this type of broadcast two separate channels are transmitted on the same carrier using FM. The information is encoded via four separate sets of signals.

QUESTIONS

3.5 (a) In the MF band the allowed range for radio broadcasts is 526 to 1606 kHz. How many 9 kHz channels are available?

(b) If the carrier frequency is 1 MHz and the maximum modulating frequency is 4.5 kHz, calculate the maximum and minimum frequencies of the transmitted AM wave.

(c) A carrier of 1053 kHz is amplitude modulated with signal frequencies in the range 50 Hz to 4.5 kHz. Calculate the actual range of frequencies transmitted and the bandwidth.

3.6 Sketch the resultant waveform that would be sent if a 1 kHz square wave were to be transmitted using a 4.5 kHz bandwidth.

3.4 THE PROPAGATION OF RADIO WAVES

The ionosphere

Although radio transmissions were known to be transmitted as waves, early experimentation relied on the premise that the waves travelled in straight lines. In 1901 Marconi received a brief radio message in Newfoundland transmitted across the Atlantic Ocean from Cornwall. Not only did this bring in the era of long distance radio communication, it showed that the propagation of radio waves was more complex than originally thought. As Fig 3.12 shows, the curvature of the Earth prevents straight line transmission between the two places. The solution to this problem was provided by Edward Appleton. He found that there was a 'reflecting' layer high above the surface of the Earth. It was called the **ionosphere**.

Fig 3.11 Preparing for Marconi's radio transmission in 1901.

Newfoundland   Ireland

Fig 3.12   The effect of the curvature of the Earth.

After Appleton's initial discovery, further work showed that the ionosphere did not consist of just a single layer. It consisted of a number of layers and their effects on radio waves depend on various factors. The ionosphere is due to high frequency electromagnetic radiation (mainly ultraviolet and X-ray) from the Sun. The solar radiation ionises the gases (mainly nitrogen and oxygen) in the upper layers of the atmosphere into positive ions and free electrons. Although recombination occurs between individual positive ions and electrons, there are usually large numbers of charged particles present. The greater the number of charged particles the greater the reflection effects.

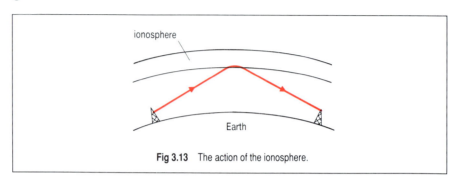

Fig 3.13   The action of the ionosphere.

INVESTIGATION

The wave properties of radio waves

The wavelengths of radio waves used for communication purposes are such that it is not easy to demonstrate their wave properties in the laboratory. However it is possible to illustrate these properties with the type of radio waves called microwaves. With the apparatus used, the frequency is 10.7 GHz and so the wavelength is 0.028 m. Although these frequencies are similar to those used in microwave cookers, the power of the waves is so low that there is no danger.

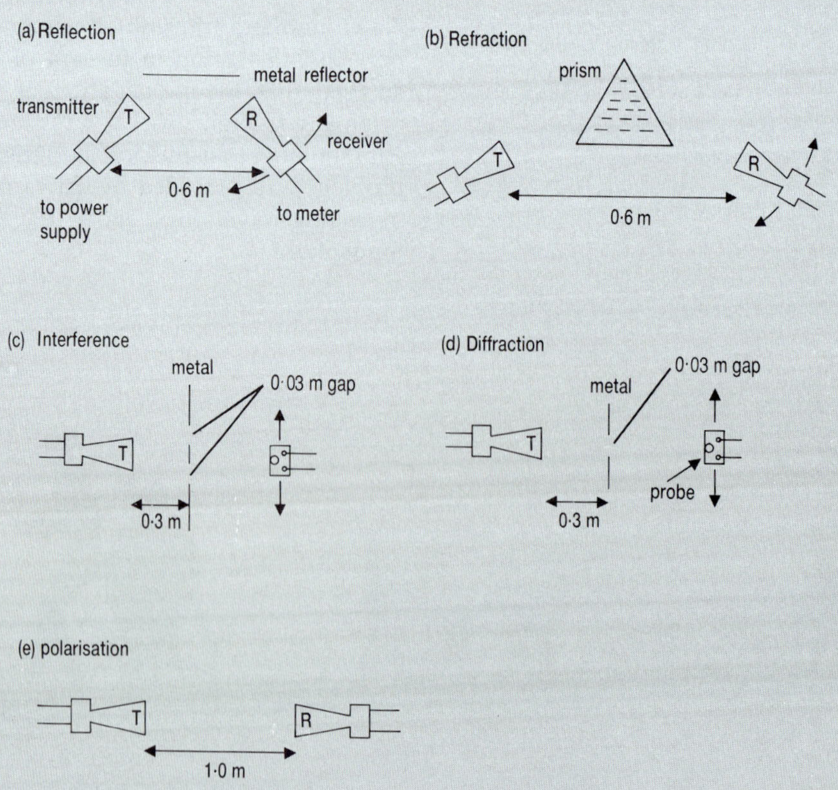

(a) Reflection

metal reflector

transmitter

T   R   receiver

0·6 m

to power
supply   to meter

(b) Refraction

prism

T   R

0·6 m

(c) Interference

metal   0·03 m gap

T

0·3 m

(d) Diffraction

metal   0·03 m gap

T

probe

0·3 m

(e) polarisation

T   R

1·0 m

Fig 3.14   The wave properties of microwaves.

TRANSMITTING RADIO WAVES

You will need: a microwave kit – this should include the transmitter, receiver, power supply, diode probe, appropriate meter for receiver and probe, metal sheets and stands, hollow prism filled with paraffin oil, polarisation grill; low voltage power supply.

1. Set up the transmitter and receiver. Follow the manufacturer's instructions to check their operation.

2. To demonstrate reflection set up the apparatus as shown in Fig 3.14(a). Does the angle of incidence equal the angle of reflection?

3. To demonstrate refraction set up the apparatus as shown in Fig 3.14(b). Trace out the path of the microwaves to check how the waves change direction when they enter and leave the liquid.

4. To demonstrate interference set up the apparatus as shown in Fig 3.14(c). Move the diode probe as shown. The probe will detect a number of positions of constructive and destructive interference.

5. To demonstrate diffraction set up the apparatus as shown in Fig 3.14(d). Why do you think the gap was set to 0.03 m? Move the diode probe as shown. Does diffraction occur?

   Due to diffraction the angle θ at which the first minimum is detected is given by the following expression:

   $$\sin \theta = \lambda/d \quad \text{where } \lambda = \text{wavelength}$$
   $$d = \text{width of the slit}$$

   Using suitable values of d, check the accuracy of this formula.

6. To demonstrate polarisation, set up the apparatus as shown in Fig 3.14(e), with transmitter and receiver upright. The waves from the transmitter are then vertically polarised. To detect the signals the receiver must be in the same plane. If the receiver is rotated through 90° the signal power drops to zero. Investigate the effect of placing the grill between the transmitter and receiver. Explain your results.

Radio waves and the ionosphere

The action of these layers of ions on the travelling radio waves is complex although the resultant effect is fairly easy to understand. As the waves pass into the ionosphere they undergo a series of refractions as they pass through the different layers. The speed of the waves increases and they are refracted away from the normal until eventually total internal reflection occurs (this phenomenon is explained in more detail in Section 6.2. The total effect of the ionosphere is thus one of reflection – just like the metal sheet reflecting the microwaves in the last investigation. This is a simple explanation. What actually happens to the waves as they pass into the ionosphere depends on a number of factors.

- The amount of solar radiation falling on that part of the Earth affects the number of charged particles present. This is not constant varying over the year and with the amount of sunspot activity. In the winter less solar radiation falls on the area of the ionosphere above the winter zone. Also the larger the number of sunspots the greater is the amount of solar radiation that falls on the Earth. The Sun goes through an eleven year cycle of sunspot activity.

- The time of day. At night recombination occurs in the ionosphere removing many of the ions. Also the heights of the various layers change during the 24 hour period.

• The frequency of the radio transmissions. As the frequency of transmission increases, the effectiveness of the ionosphere as a reflector decreases. This means more and more of the radiation passes through the ionosphere without being reflected back to Earth. The upper frequency limit is usually taken as 30 MHz (i.e. above this frequency all the radiation will pass through the ionosphere). But, depending on the conditions, the upper limit can be as low as 20 MHz or as high as 100 MHz.

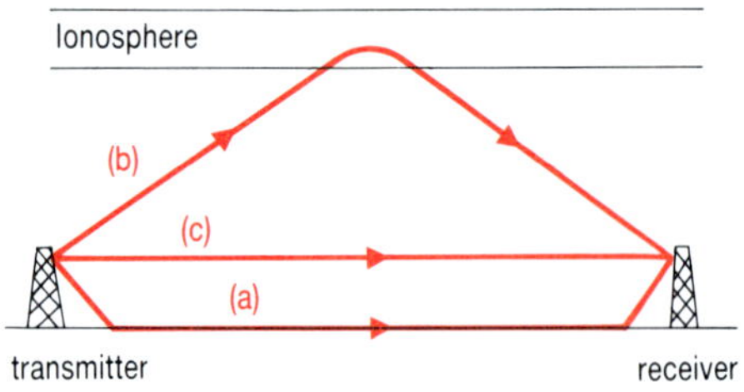

How radio waves travel

Although the ionosphere plays an important part in radio wave propagation it is not the only method. There are three main methods of propagation altogether and radio waves can often reach the receiver using more than one route.

Ground (or surface) waves These follow the surface of the Earth. The range of these signals depends on how good the surface is as a conductor. The better the conductor the further the wave will travel. For example, radio transmissions will travel further over a water surface than they will over sand. Why is water a better conductor than sand? The maximum distance for this method of propagation is of the order of 1000 km. This is the most important method of propagation for waves up to 2 MHz. Because the waves travel along the ground they are not affected by changes in the ionosphere. Above 3 MHz, the higher the frequency of the waves the more rapidly they are attenuated.

Sky waves This is the most important method of propagation for transmissions in the range 3 to 30 MHz. With such frequencies it is possible to establish world-wide communications. The wave can travel around the Earth's surface via a series of reflections between the two reflecting surfaces of the ionosphere and the ground. The ground acts as a conductor because it contains charge carriers and these have a similar effect to the ions in the ionosphere on the waves.

Space waves For frequencies above 30 MHz, transmissions generally pass through the ionosphere. Ignoring satellite communication for the moment (this will be discussed in detail in Chapter 5), it means that only waves passing from transmitter to receiver in a straight line can be detected. This is often called 'line of sight' transmission. The range will depend on the height of the transmitter and receiver together with the presence of any intervening obstacles. The upper limit to line of sight transmission is often taken as 100 km though it can be further under suitable conditions.

Fig 3.15 The propagation of radio waves (a) ground waves (b) sky waves (c) space waves.

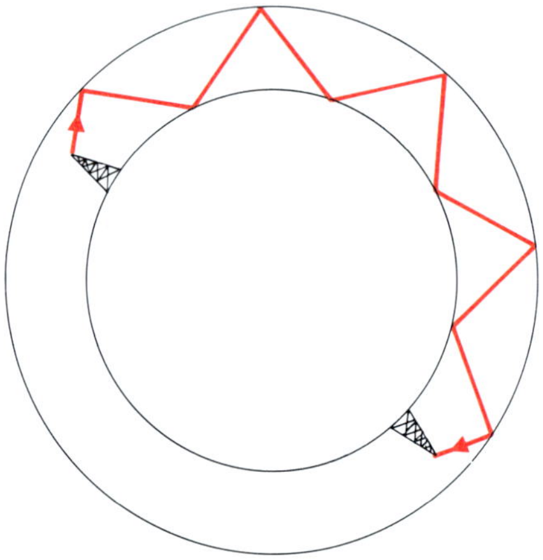

Fig 3.16 Reaching the other side of the Earth.

QUESTIONS

3.7 Which would normally be the main method of propagation (as ground, sky or space waves) for the following broadcasts:
(a) Radio 4 at 200 kHz,
(b) Radio 2 at 909 kHz,
(c) Radio 3 at 93.5 MHz?

3.8 Use the wave properties of radio waves to explain the following:
(a) Two images, one much weaker than the other, sometimes appear on a television screen when an aircraft flies overhead.
(b) Even in the 'shadow' of buildings it is still possible to receive radio broadcasts.
(c) If radio waves can travel by more than one path to the receiver fading of the received signal sometimes occurs.
(d) In the summer it is sometimes possible to receive TV broadcasts from countries in Europe.

TRANSMITTING RADIO WAVES

Adapted from *Electronic Telecommunication Systems* by Frank Dungan

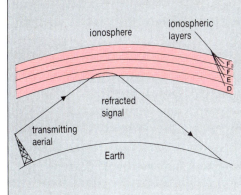

Fig 3.17 Sky wave propagation.

The ionosphere

The behaviour of the sky wave is a result of the ionosphere. There are different degrees of ionisation, forming several recognisable layers. An illustration of the ionospheric layers is shown in Fig 3.17.

The D layer, 50–90 km above the Earth, is the lowest layer and exists only in the daytime. Since this layer is furthest from the Sun, its ionisation is relatively weak; consequently it does not affect the travel direction of radio waves. However, the D layer does absorb energy from the electromagnetic wave and, when present, attenuates the sky wave in both directions – on the way up to the other layers and on the way back down. Signals in the MF band are completely absorbed by the D layer, thereby limiting these signals to ground wave propagation during daylight hours. At night, however, the D layer disappears, and long distance MF transmissions via sky wave propagation are possible.

The E layer, 90–150 km above the Earth's surface, reaches its maximum density at noon. Its ionisation is so weak at night that the layer may disappear.

The F layer, 150–400 km above the Earth's surface, splits into F_1 and F_2 in the daytime, with F_2 varying from summer to winter. The F_1 layer ranges from about 150 to 250 km above the surface of the Earth. The F_2 layer is closest to the Sun and ranges in height from about 150 to 300 km on a winter day, and from about 300 to 400 km on a summer day.

1. Explain what happens to the radio waves as they pass into the ionosphere.

2. How are the ions formed in the ionosphere?

3. Why is the ionosphere so important in radio wave propagation?

4. Why does the D layer exist only in daytime?

5. What is the difference between daytime and night-time propagation for MF waves?

6. Why does the E layer reach its maximum density at noon?

7. Why is there a difference in height of the F_2 layer in summer and winter?

SUMMARY

Electromagnetic waves are transverse electrical and magnetic waves travelling in phase. Radio waves are very important in telecommunications and form part of the electromagnetic spectrum. The first radio transmissions used carrier keying for sending messages. Radio waves of suitable frequency can travel long distances. Nowadays the audio frequency signals are encoded onto a radio frequency carrier. Encoding of the signals onto the radio carrier is done using amplitude or frequency modulation. The ionosphere plays an important part in radio transmission though radio waves reach the receiver as ground waves, sky waves or space waves.

3.9 **(a)** Explain the meaning of the following terms: wavelength, amplitude and phase.
(b) Why is modulation essential for radio broadcasting?

3.10 In this question assume that the channel bandwidth of an AM signal transmitted on radio waves is 9 kHz. Explain the following statements:
(a) The human voice sounds relatively undistorted.
(b) Some types of musical instruments can sound distorted.
(c) A transmitted 3 kHz square wave would be received as a sine wave.

3.11 **(a)** Describe, with the aid of diagrams, how you could demonstrate experimentally the waveform resulting from the addition of two sinusoidal waveforms of different frequency and amplitude.
   Explain briefly the significance of waveform addition to radio-transmission by amplitude modulated waves.
(b) What physical principles underline the simultaneous transmission of messages along a single link? (COSSEC, AS, spec.)

3.12 What is meant by the terms
(a) bandwidth,
(b) carrier wave, and
(c) amplitude modulation?
A carrier wave of frequency 18.0 kHz is modulated by an audio signal containing a continuous spectrum of frequencies between 300 Hz and 3400 Hz. Sketch the frequency spectrum of the modulated signal, indicating the upper and lower side bands. (ULSEB 1987)

3.13 **(a)** What do you understand by the term *amplitude-modulated carrier wave*?
   Give a labelled block diagram showing the basic elements of a simple amplitude-modulated radio transmitter for the broadcasting of audio signals developed by a microphone.
(b) Speech signals in the frequency range 300 Hz to 3400 Hz are used to amplitude-modulate a carrier wave of frequency 200.0 kHz. Determine
   (i)   the bandwidth of the resultant modulated signals,
   (ii) the frequency range of the lower sideband,
   (iii) the frequency range of the upper sideband.

   Fig 3.18 shows the range of speech signals and the carrier wave. Copy the diagram into your answer book and use it to represent your answers to (i), (ii) and (iii). (ULSEB 1989)

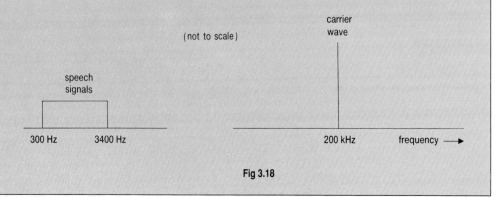

Fig 3.18

3.14 From the dimensions of the transmitter on Hertz's original apparatus (illustrated in Fig 1.12) the wavelength of the radio waves was estimated to be of the order of 5 m. Describe how you would demonstrate their wave properties assuming you had his original apparatus. In your answer include the dimensions of any other apparatus you would use.

Chapter 4

RECEIVING RADIO WAVES

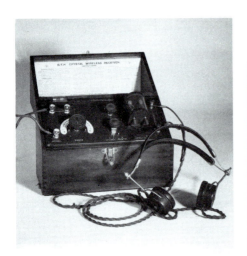

Fig 4.1   An early crystal receiver.

LEARNING OBJECTIVES:

After studying this chapter, you should be able to:

1. understand the action of inductors and capacitors in a.c. circuits;

2. describe the use of the parallel inductor-capacitor circuit as a tuner for a radio receiver;

3. know the basic circuit for a crystal set receiver;

4. recall improvements that can be made to the basic crystal set to improve its sensitivity and selectivity;

5. recall the action of the ferrite rod aerial in LF and MF wavebands;

6. recall the action of the dipole aerial and how to improve its reception characteristics;

7. know about other uses of the radio spectrum apart from broadcasting.

4.1 RECEIVING THE REQUIRED SIGNALS

Early receivers were difficult to operate and the aerial was of extreme importance. Receivers operated over only a small range of frequencies. Nowadays receivers are highly sensitive and can be made to respond to a wide range of frequencies. Most people have receivers for picking up the broadcasting services in their homes and often in their cars. Mobile phones are becoming an essential part of the business person's equipment. Radio links have become an essential communication link for many organisations and services. Most of this chapter will be concerned with the reception of radio broadcasts.

The inductor and capacitor

An **inductor** consists of a coil of wire that is often wound over a magnetic core. The core is usually a ferrite rod. **Ferrite** is a material that can easily be magnetised but is an electrical insulator (unlike iron). If the current through an inductor does not change then it behaves just like a resistor. If the current

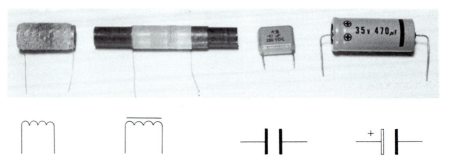

Fig 4.2   Inductors and capacitors.

changes then electromagnetic induction opposes the change. This effect is of importance in a.c. circuits where the currents are continually changing. For more details on the action of the inductor consult a basic physics book such as the core book *Physics*, Chapter 19.

The simplest **capacitor** consists of two metal plates separated by an insulator such as air, paper or polythene. When charged, one plate is charged positively and the other negatively. In a d.c. circuit the current drops to zero when the plates are charged up. In an a.c. circuit the plates are continually being charged, discharged and charged up the opposite way round. For more details on the action of the capacitor consult a basic physics book such as the core book *Physics*, Chapter 17.

Reactance

In a d.c. circuit, the ratio of voltage to current is called the resistance. In an a.c. circuit, this ratio is called the **impedance** Z. It is calculated in the same way but the values of p.d. and current must be either both r.m.s. (root mean square) or both peak values. For a resistor the value is the same in d.c. or a.c. circuits. With a resistor in an a.c. circuit the voltage and current are in phase. This is not the case with either the inductor or capacitor. With the inductor the current lags the voltage by 90°. This means that the current reaches its maximum value one quarter of a cycle after the voltage. With the capacitor, the current leads the voltage by 90°. These effects are illustrated in Fig 4.3. For more details on these effects consult a basic physics book such as the core book *Physics*, Chapter 20.

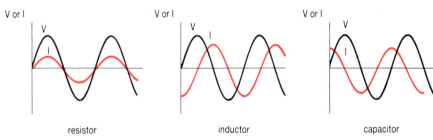

Fig 4.3   Phase relationships in a.c. circuits.

Due to the phase differences between current and voltage there is no actual power loss in a perfect inductor (having zero resistance) or a perfect capacitor (having infinite resistance), unlike a resistor, and the a.c. impedance is called the **reactance** X. The reactances are given by the following formulae:

Reactance of an inductor: $X_L = 2\pi fL$ where f = frequency
L = inductance (units: henry, H)

Reactance of a capacitor: $X_C = \dfrac{1}{2\pi fC}$ where f = frequency
C = capacitance (units: farads, F)

From these equations it can be seen that the reactance of an inductor increases with frequency, while that of a capacitor decreases. This is illustrated in Fig 4.4 for the three types of component.

The parallel inductor-capacitor circuit

The number of radio transmissions that can be picked up by a receiving aerial is quite large. Each of the transmissions that can be received produces a.c. currents corresponding to the frequencies in the carrier waves. It is then necessary to separate the required frequency from all the others before demodulation occurs.

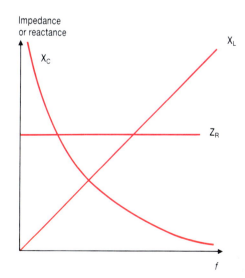

Fig 4.4   Variation of impedance/reactance with frequency.

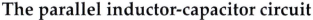

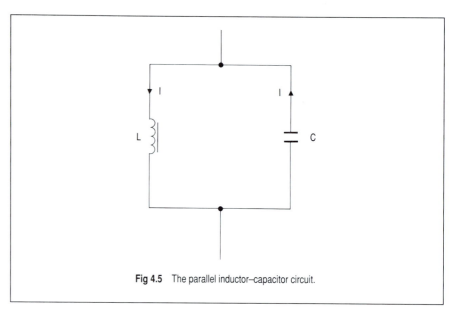

Fig 4.5   The parallel inductor–capacitor circuit.

The simplest method is the **'parallel L–C circuit'**. This consists of an inductor (L) and a capacitor (C) in parallel. At one particular frequency called the **resonant frequency** the reactances are equal. At this frequency, since the voltages and reactances are equal, the currents must be equal. However, there is a phase difference of 180° between the currents in the two branches. At one instant the direction of current flow is as shown in Fig 4.5 and, half a cycle later, the current flow is reversed. At this frequency the largest currents flow around the L–C circuit and produce a larger potential difference across the parallel components than at any other frequency. The p.d. that appears across the two components varies with frequency in the way shown in Fig 4.6 and is a maximum at the resonant frequency. If a range of frequencies is present then the L–C circuit will resonate to one particular frequency (in practice this turns out to be a narrow band of frequencies).

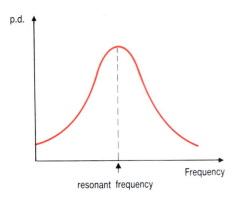

Fig 4.6   Variation of p.d. with frequency of an L–C circuit.

INVESTIGATION

Inductors and capacitors in a.c. circuits

The first part of this investigation will enable you to find out about the effects of a.c. circuits on inductors and capacitors. The second part will be done with values of inductance and capacitance that produce a resonant frequency well within the range of the usual laboratory signal generator. These resonant frequencies are much smaller than those used in radio frequency transmissions, but the principle is the same.

You will need: a signal generator;
  oscilloscope;
  a.c. ammeter;
  1 μF, 0.1 μF and 1 nF capacitors (Values are not critical but do not use electrolytic capacitors);
  1100 turn coil fitted with C-cores (its inductance is 1 H);
  1 k, 4.7 k, 10 k and 100 kΩ resistors.

1. To investigate the variation in reactance with frequency for an inductor (or capacitor) set up the circuit shown in Fig 4.7(a) with either the inductor or the capacitor. Measure the peak-to-peak p.d. across the component. With the a.c. ammeter the reactance of the inductor or capacitor can be calculated. Note that the ammeter will read the r.m.s. value of the current. The peak-to-peak value of the p.d. must be divided by 2.8 to get the r.m.s. value. Calculate the reactance at various frequencies. Do your results agree with the

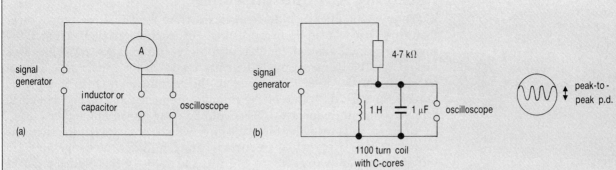

Fig 4.7 **(a)** Circuit for investigating reactance.
(b) Circuit for the measurement of the resonant frequency.

values calculated using the formula? The impedance of a resistor should not change with frequency. You could check this.

2. Set up the circuit with both the inductor and capacitor as shown in Fig 4.7(b). The resistor and the L–C circuit act as a potential divider across the input from the signal generator. At the point of resonance the p.d. across the L–C circuit will be a maximum.

3. Increase the signal generator frequency in steps of 10 Hz from 50 to 250 Hz (if using the values shown), measuring the peak-to-peak p.d. across the L–C circuit each time. From a graph of the frequency against the p.d. determine the resonant frequency of the L–C circuit.

4. You could investigate the following:
   (a) Repeating the exercise using the 1100 turn coil without the C-cores and, in turn, with the other capacitors. The resonant frequencies will be much higher in all cases.
   (b) Investigating the effect on the shape of the resonance curve of placing each of the resistors in parallel with the L–C circuit. It would be instructive to plot all the results on one graph as these will be useful in Section 4.2. What happens to **(i)** the value of the resonant frequency, **(ii)** the shape of the resonance curve?

Calculating the resonant frequency

Resonance occurs when the reactances of the inductor and capacitor are equal. At this point:

$$2\pi fL = \frac{1}{2\pi fC}$$

Therefore $f^2 = \dfrac{1}{4\pi^2 LC}$

and $\qquad f = \dfrac{1}{2\pi \sqrt{LC}}$ $\qquad$ where L = inductance, unit: henry, H
$\qquad\qquad\qquad\qquad\qquad\qquad\qquad\qquad$ C = capacitance, unit: farad, F

For example if $L = 20$ mH and $C = 0.1$ μF then the resonant frequency is given by:

$$f = \frac{1}{2\pi \sqrt{20 \times 10^{-3} \times 0.1 \times 10^{-6}}}$$

$$= \frac{1}{2\pi \sqrt{2 \times 10^{-9}}}$$

$$= \frac{1}{2\pi\ 4.5 \times 10^{-5}}$$

$$= 3.5 \times 10^3\ \text{Hz}$$

Receiving AM transmissions

In the previous investigation frequencies were used that are much lower than those used for radio broadcasting. For example, in the MF band, the frequency range allotted for this purpose is 526 kHz to 1606 kHz. This means that the values of the inductance L and/or the capacitance C must be much smaller than those used in the investigation. Also it must be possible to vary the value of the resonant frequency of the L–C circuit to receive different frequencies. This could be done by varying either L or C. In practice it is usually the capacitor which is changed to vary the frequency over a particular frequency band. In order to change between the different bands (e.g. MF to LF) it is the value of the inductor that is usually changed.

QUESTIONS

4.1 Hertz's original receiver was not very sensitive and this meant that radio waves could only be detected at a range of a few metres. A device called the coherer enabled the transmission distance to be greatly increased. It is illustrated in Fig 4.8 together with a typical receiving circuit of that period. Normally the nickel and silver filings in the coherer were loosely spaced and the resistance between the silver plugs was large. An incoming radio signal made the filings cling together and their resistance became much smaller.

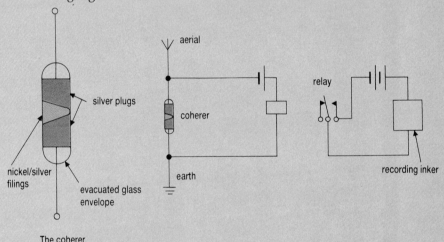

The coherer

Fig 4.8 An early receiving circuit.

(a) Using Fig 4.8 explain how the receiver worked.

(b) In practice the coherer required a physical 'tap' to break up the filings and return it to its high resistance state. How would you redesign the circuit to do this automatically? You will need a component to 'tap' the coherer.

(c) If you had been given the task of improving the sensitivity of the coherer what factors would you try investigating to achieve this?

4.2 (a) Calculate the reactance of a 0.1 H inductor at 1 kHz. At what frequency would the reactance of a 10 nF capacitor be equal to this?

(b) If $L = 1.5$ mH and $C = 1$ μF, calculate a value for the resonant frequency.

(c) Given an inductor of value 0.1 H, what value of capacitance would be needed for a resonant frequency of 1 kHz?

(d) Given a capacitor of value 50 nF, what value of inductor would be needed for a resonant frequency of 10 kHz?

(e) In a radio receiver, the value of the inductor is 20 µH and that of the capacitor is 470 pF. Calculate the resonant frequency.

(f) Calculate the resonant frequencies of the L–C circuits that you used in the previous investigation. (The inductance of the 1100 turn coil with C-cores is about 1.0 H and without them is 20 µH). Try to account for any differences between the calculated values and the experimentally determined values.

4.2 DEMODULATING THE SIGNALS

In the years immediately preceding World War II the BBC became very concerned that their transmitters could act as beacons for allowing enemy aircraft to obtain bearings on their position. Their solution, put into effect immediately the war started, was that all transmitters broadcasting the same programme used an identical frequency. Also if an aircraft was found to be approaching a particular transmitter it was switched off.

Receiving the signals

A parallel L–C circuit connected to an aerial and an earth is shown in Fig. 4.9. This is called the **tuning circuit**. Incoming radio waves induce alternating currents in the aerial. The purpose of the earth is to allow any currents generated by the 'unwanted' frequencies to escape. The variable capacitor allows the value of the resonant frequency to be changed. At the resonant frequency the largest p.d. is produced and this is applied across the rest of the receiver.

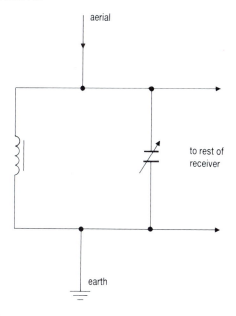

Fig 4.9 The first stage of a receiver.

INVESTIGATION

'Looking at' r.f. transmissions

In this investigation you can observe the incoming radio transmissions by displaying them on the oscilloscope using a very fast timebase. You can use a homemade tuning circuit or a manufactured module containing a tuning circuit.

You will need: a ferrite rod(for the homemade device);
insulated wire;
paper for winding the coil on;
sticky tape;
variable capacitor (up to 600 pF),
or a tuning circuit module;
oscilloscope.

1. Building the homemade tuning circuit: Make a paper cylinder to fit onto the ferrite rod. Wind 50 turns of the wire onto the cylinder.

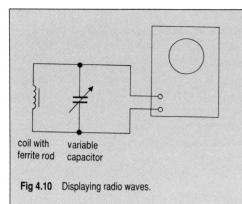

coil with   variable
ferrite rod capacitor

Fig 4.10   Displaying radio waves.

Remove the insulation from the ends and secure them with the tape to the cylinder. An aerial and earth are not required here.

2.   With the circuit shown in Fig 4.10 set the oscilloscope to its fastest timebase speed and the amplification to its maximum.

3.   Adjust the variable capacitor until a strong (i.e. large amplitude) signal is observed on the screen. It may be necessary to change the orientation of the ferrite rod to maximise the signal. If the timebase speed is then slowed down it is possible to see the changes in amplitude of the carrier due to AM. Note that these changes will vary continuously.

4.   (a)   Measure the frequency of the carrier wave. How would you find out which broadcast you were receiving?
     (b)   Measure the peak-to-peak voltage of the received carrier. What is the amplitude of the carrier?
     (c)   Vary the resonant frequency of your circuit and find out how many stations you can receive.

With the radio receivers purchased in the early 1920s it was necessary to purchase a licence for their operation (rather like television today). The user was allowed to install an aerial up to 30 m in length – and all of this was needed unless the receiver was very close to the transmitter.

The simplest type of receiver

In this type of receiver there is no amplification of the received signals. The voltage developed across the L–C circuit is passed to the demodulator (often called the detector) to decode the required information from it. This consists of a **semiconductor diode**. A diode allows current to flow through it in one direction only, as with the valve diode. This means that if an a.c. signal is applied to it then only half the signal passes through it. This is called **rectification**. The receiver is still usually called a crystal set as the demodulator used in the original sets was a crystal of the mineral galena and this acted in a similar way to the diode used today. A germanium diode rather than the more usual silicon diode is used here. It has an important advantage with radio transmissions. The germanium diode requires a smaller p.d. across it before it conducts (about 0.2 V compared with silicon's 0.6 V) and this is very important in this kind of receiver where the signal strength detected is small and produces only small p.ds across the tuning circuit. Fig 4.11 shows what happens to the signal as it passes through the diode. The rectified signal contains half the modulated carrier.

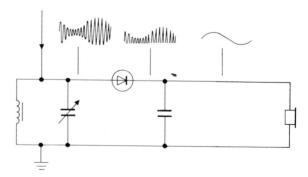

Fig 4.11   A simple receiver.

An earpiece is attached between the diode and the earth line. However, reception is improved if a capacitor is connected in parallel as shown in Fig 4.11. The value of the capacitor is chosen so that it provides a low impedance path for the r.f. component of the signal but a high impedance path for the a.f. component. The a.f. part of the signal is then passed onto the earpiece. The earpiece must have a high impedance. You will remember from the investigation in Section 4.1 that putting a low value resistor in parallel with the L–C circuit reduces the p.d. produced across it. The same thing happens here. A low resistance earpiece will reduce the p.d. and the resultant sound signal will be smaller in amplitude.

Building a 'crystal' set

The original crystal was the mineral galena (this is composed mainly of lead sulphide) attached to a 'cats whisker' (just a wire). By moving the wire over the surface of the crystal it was possible to find a position at which rectification of the incoming carrier occurred. However, the conventional germanium diode will be used here. Do not expect superb reception using this circuit but console yourself with the fact that it does not require any external source of power. It is designed for the reception of transmissions in the MF band.

You will need: a tuning circuit such as that constructed for the
previous investigation;
germanium diode;
1 nC capacitor;
high impedance earpiece;
a long length of wire for the aerial;
oscilloscope.

1. The aerial should be vertical and as long as possible. For the best results the aerial should be outside the building and on the side closest to the nearest MF band transmitter. The earth is very important. It is simplest to use the external earth connection on an oscilloscope or a low voltage power supply. Never make the earth connection via the earth of a mains socket. The receiver is set up as shown in Fig 4.11.

Table 4.1 Broadcasting frequencies for Radio's 1 to 4

Radio	1	2	3	4
frequencies/ kHz	1053 1089	693 909	1215	720

2. You may only be able to detect one station or even none at all. If nothing is detected then build the improved receiver described in the next investigation. It is useful to know what station you are listening to. Table 4.1 shows the frequencies of BBC's Radios 1 to 4 that are nationally broadcast. You will probably have one or more local stations also broadcasting on the MF band. Consult the *Radio Times* to find further details of BBC local radio. Also, you may have an independent station in your area.

3. With an oscilloscope the change in the received signal as it passes through the circuit can be investigated.

Improving the output

With your set you may only have been able to receive one station or even none at all. One cause of problems is the aerial and another is the absence of amplification of the received signal. We want a radio that has both sensitivity and selectivity. **Sensitivity** refers to the ability of the receiver to detect 'weak' signals. With the crystal set there was no amplification of the received signal and the weaker the signal the less likely the crystal set would detect it. **Selectivity** refers to the ability to distinguish between transmissions that are close together in frequency. With the crystal set you will have noticed that one station could be received over quite a large tuning range or even two stations could be detected simultaneously. The selectivity is affected by the shape of the resonant curve for the L–C tuning circuit (as shown in Fig 4.12). The flatter the curve the less selective the

Radio receivers in use today invariably use the superheterodyne system. In this type of receiver the incoming r.f. signal is mixed with an r.f. signal generated within the receiver to give what is called an intermediate frequency (usually 455 kHz). The rest of the receiver is designed to optimise performance at this particular frequency.

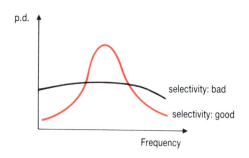

Fig 4.12   Selectivity of a tuning circuit.

receiver is. The following investigation suggests two ways in which modest improvements could be made to the simple receiver.

INVESTIGATION

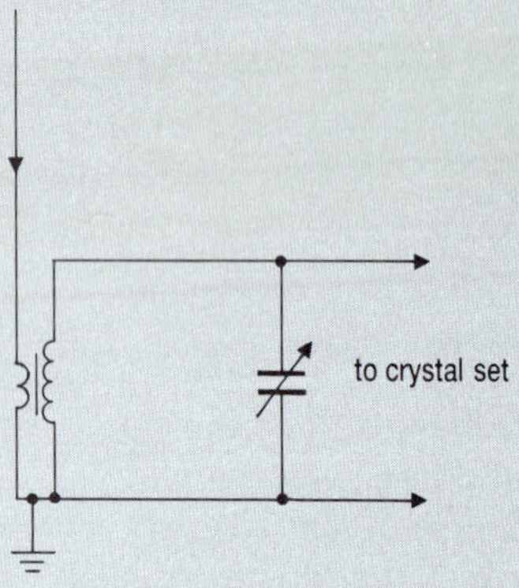

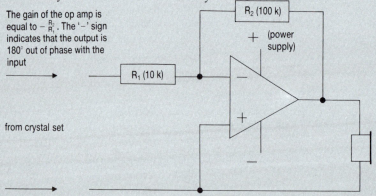

Fig 4.13   A more efficient aerial.

Fig 4.14   Amplifying the signal.

Improving the basic receiver

You will need: the same apparatus as for the previous investigation; an operational amplifier module and its power supply.

1. The efficiency of the aerial can be improved if the coil of the tuning circuit is separate from the aerial itself. This is easily done by winding about 10 turns of wire around the coil already constructed. One end of the 10 turn coil is then connected to the aerial and the other to the earth as shown in Fig 4.13. This will improve the selectivity but reduce the sensitivity.

    The gain of the op amp is equal to $-\frac{R_2}{R_1}$. The '−' sign indicates that the output is 180° out of phase with the input

2. The received signal needs to be amplified. One of the simplest ways is to feed the signal, after detection, to the input of an **operational amplifier** (op amp). This, as the name suggests, will amplify the signal. Although the op amp can be wired up using a breadboard or similar apparatus, the use of an op amp module is much easier. The circuit is shown in Fig 4.14. The op amp is connected up to act as an inverting amplifier. This means that the output is 180° out of phase with the input but this will make no difference to the sounds heard.

3. It is hoped that both these suggestions will improve the number of stations received and their strength. You could try changes in the two suggestions (i.e. changing the number of turns on the aerial coil and varying the gain of the op amp) to see if any further improvements are produced.

4. (a) If you can now receive a number of stations you could make a calibrated scale to go onto the tuning knob of the variable capacitor.

    (b) If you construct a coil with 200 turns instead of 50 then you should be able to receive LF band transmissions. Radio 4 is broadcast at 198 kHz in this band.

A practical MF band receiver

Radio reception can be greatly improved if the incoming r.f. signal is amplified before being applied to the detector. This type of amplifier is referred to as an r.f. amplifier and the complete receiver is termed a **tuned radio frequency (TRF) receiver**. Why do you think better reception is obtained if the incoming signal is amplified before it passes through the diode? In most AM receivers there is no need for an external wire aerial as the ferrite rod on which the coils are wound is used as the aerial. This responds to the magnetic component of the incoming electromagnetic wave and is considered further in Section 4.3.

One way of building a TRF receiver is to use an **integrated circuit** (IC). An IC is an electronic circuit built on a tiny piece of silicon, such as the op amp in the last investigation. It can contain diodes, transistors, resistors and capacitors. A TRF IC is made by Ferranti and contains a complete r.f. amplifier, detector and automatic gain control (AGC). AGC keeps the output signal constant even if the strength of the received signals changes. The IC requires only a tuning circuit and earpiece to produce a complete AM receiver. Two versions are produced. Although the ZN414Z outwardly looks very much like a transistor it is an integrated circuit and will drive a crystal earpiece directly while the ZN416E, with its larger output, will drive a pair of headphones. To increase the signal strength sufficiently to drive a loudspeaker, the output from the TRF IC must be connected to the input of an a.f. amplifier, usually a type of op amp. Often there is more than one stage of audio amplification. With suitable tuning circuits the receivers can be used over the complete LF and MF broadcasting bands.

While it is possible to build a receiver using a ZN414Z IC, an a.f. amplifier and a few external components, it is quicker and easier to build it using an electronic systems kit. One of these is illustrated in Fig 4.16 and it uses the ZN414Z IC. If you have such a kit then you can make an excellent MF band radio in just a few minutes. It can be tuned over the frequency range 500 kHz to 1600 kHz.

Fig 4.15 The ZN414 Z and ZN416 E.

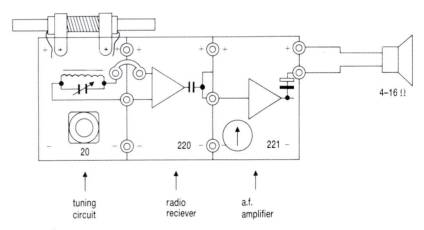

Fig 4.16 A practical MF receiver.

QUESTIONS

4.3 (a) In the first investigation in Chapter 3, try to explain why 'sounds' were heard although there was no apparent modulation of the carrier wave.
(b) Where does the energy required to drive the earpiece in the crystal set come from?
(c) Explain why MF band reception is usually better at night.
(d) Fig 4.6 shows the variation in p.d. across a tuning circuit with frequency. How would the shape of the curve change if the circuit were (i) less sensitive and (ii) more selective? What would

4.3 AERIALS

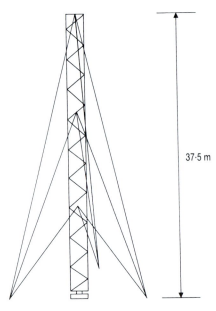

Fig 4.17 A transmitting aerial.

It was mentioned in Chapter 2 that it was not practical to change sound waves into electromagnetic waves of the same frequencies for propagation. Apart from the fact that only one broadcast could take place in a particular area, broadcasting at a.f. frequencies would mean that the aerial would have to be at least 7.5 km long. Also, large amounts of power would be required to send the waves any significant distance.

Transmitting and receiving aerials

An **aerial** is a device for transmitting or receiving radio waves. The word **antenna** is increasingly being used in place of aerial and you may find this in your reading. The function of the aerial is to convert electrical energy into electromagnetic wave energy or vice versa. With a transmitting aerial, the aim is to produce as much electromagnetic radiation as possible from the r.f. currents flowing in the aerial. At the receiving aerial, the aim is to collect as much electromagnetic radiation associated with the required signal as possible. Strictly speaking there is little difference between transmitting and receiving aerials and sometimes the same aerial is used for both (e.g. with mobile radio links).

The main difference between the transmitting and receiving aerials used in radio broadcasting is in the quantity of power involved. Radio 4 in the LF band is broadcast at a power of 400 kW and the receiving aerial will be detecting a signal with a power of only a few milliwatts. Also the transmitting aerial is usually designed to radiate the electromagnetic waves in all horizontal directions (i.e. it is non-directional) while the receiving aerial usually receives signals from a fixed direction. With AM broadcasts the aerial is usually vertical and the transmitted signal is vertically polarised.

Aerials are most efficient at converting the energy when they are 'tuned' to the particular frequency involved. This involves resonance and is most important with transmitting aerials when they are broadcasting at a particular fixed frequency. One arrangement is for the aerial to be equal in length to half the wavelength of the transmitted wave. At lower frequencies this would make the aerial impossibly long and it is possible to use one that is a quarter of a wavelength. With the ground acting as a reflector it behaves as though it were half a wavelength long. It would appear necessary to vary the length of the aerial if the transmitted frequency changes, however external electrical changes can be made that effectively do the same thing. When broadcasting from one fixed point to another it is often necessary to beam the energy from transmitter to receiver.

Picking up the transmissions

The received signal must have sufficient power if it is to be satisfactorily decoded. If the waves were emitted from a point source (this is not a practical possibility for radio waves) then the power carried by the waves would decrease with the square of the distance. This means that, as the distance is doubled, the power carried by the waves decreases by a factor of four. This is illustrated in Fig 4.18. As the distance is doubled, the total energy is spread out over an area four times as large and so the power becomes one quarter. This is called an **inverse square law** and is standard for radiations being emitted equally in all directions from a point source.

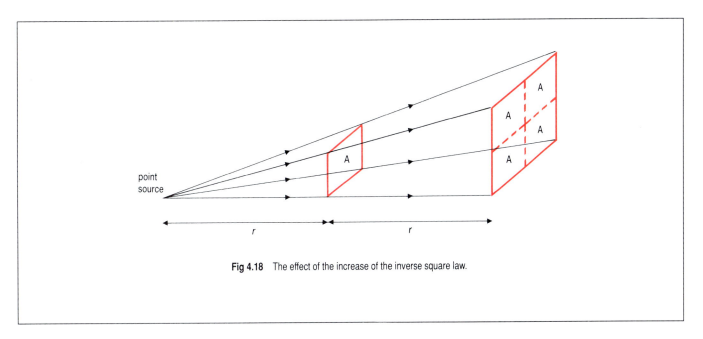

Fig 4.18   The effect of the increase of the inverse square law.

This relationship assumes that there is no absorption of radiation by the medium. In practice, there will be some absorption by the medium and electrical noise is added to the signal. Transmitting aerials obviously try to propagate the waves in the directions in which they are going to be received.

Metal aerials

When receiving radio broadcasts it is not too important for the aerial to be tuned to the frequency being received. This is fortunate as it would make aerials impossibly long except in the VHF band. Also the receiving aerial needs to be able to respond to a wide range of frequencies. The vertical metal aerial is an example of an **untuned aerial** and this means that it will respond to a wide range of frequencies. A **tuned aerial** (its construction is described later on) is designed to respond only to a set frequency (or small range of frequencies). With a radio receiver trying to detect LF and MF band transmissions, the aerial must be vertical as the transmissions are vertically polarised. It is usually a whip aerial, which is made of telescopic sections, so that it can easily be extended or retracted. The same type of aerial is often used in the VHF band with portable receivers.

Ferrite rod aerials

In practice, most radio receivers apart from car radios use a different type of aerial called a **ferrite rod aerial** for reception in the LF and MF bands. You will remember that the inductor for the L–C circuit consisted of a ferrite rod on which a coil was wound. It detects the magnetic part of the electromagnetic wave and must be orientated so as to be at right angles to the electrical component i.e. parallel to the magnetic component. If the electrical part of the wave is vertically polarised, then the ferrite rod must be horizontal and 'broadside on' to the transmitting aerial.

Fig 4.19   A ferrite rod aerial.

Tuned receiving aerials

As with transmitting aerials, receiving aerials are more efficient when they are 'tuned' to the incoming signal. The length of the receiving aerial must be equal to half the wavelength of the electromagnetic wave. For LF and MF transmissions the length of the aerial would be too large for practical purposes. For example, the 198 kHz Radio 4 transmission would need an

RECEIVING RADIO WAVES

aerial of length 750 m (wavelength $\lambda = 3 \times 10^8 / 1.98 \times 10^5 = 1.5 \times 10^3$ m). The ferrite rod aerial is a type of tuned aerial as it forms part of a tuning circuit adjusted so as to respond to a particular frequency.

The dipole aerial

At the frequencies used in the VHF (radio) and UHF (TV) bands the wavelength allows tuned aerials to be employed. At 100 MHz, the wavelength is 3 m. One of the simplest tuned aerials is the **dipole aerial**, sometimes called a Yagi aerial after its inventor. The device consists of two lengths of metal each one quarter of a wavelength long. The signal is fed to or from the aerial via a coaxial lead attached to the centre of the aerial. The aerial is mounted vertically or horizontally according to the plane of polarisation of the transmitted signal.

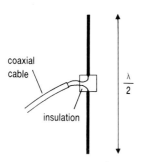

Fig 4.20 The dipole aerial.

Directional aerials

The way in which electromagnetic waves are radiated from a dipole is illustrated in Fig 4.21. These are often called **polar diagrams** and are a plot of the power of the wave in different angular directions. If the dipole transmitter is vertical, then the radiation is emitted equally in a particular horizontal plane. However, in a vertical plane, the radiation pattern is not so simple. Figure 4.21(b) shows that the maximum is in a direction through the mid-point of the aerial perpendicular to its axis.

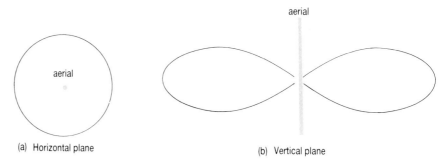

(a) Horizontal plane   (b) Vertical plane

Fig 4.21 The radiation pattern for a dipole transmitter.

Although a transmitting aerial will usually be beaming the signals in all horizontal directions the receiving aerial will be directional. The transmitting aerials for the broadcasts in a particular area are often mounted together on a tall tower. For example BBC and ITV television broadcasts will be transmitted from the same place and so the domestic receiving aerial does not need to be changed in direction. With a receiving dipole aerial it is possible to make the aerial directional and to increase the power of the signal received. This is called the gain of the aerial. The **gain** is the ratio of the signal power of the aerial received to the signal power of a simple dipole. It is usually expressed in dBs.

Figure 4.22 shows how the gain can be increased. Two extra elements are added to the basic dipole aerial – these are called a director and reflector. The director needs to be slightly shorter than the dipole (about 5 per cent less) and the reflector slightly longer than it (about 5 per cent more). The separations are expressed in terms of the wavelength in the diagram. As the names suggest, the director is placed between the dipole and the transmitter and the reflector behind the dipole. The joint effect of these two is to increase the received signal strength. If an aerial is designed to receive a range of frequencies such as 88 to 108 MHz in the VHF band then the average frequency can be used for calculation purposes. In practical aerials there is usually more than one director. More directors will increase the gain but there is an upper limit.

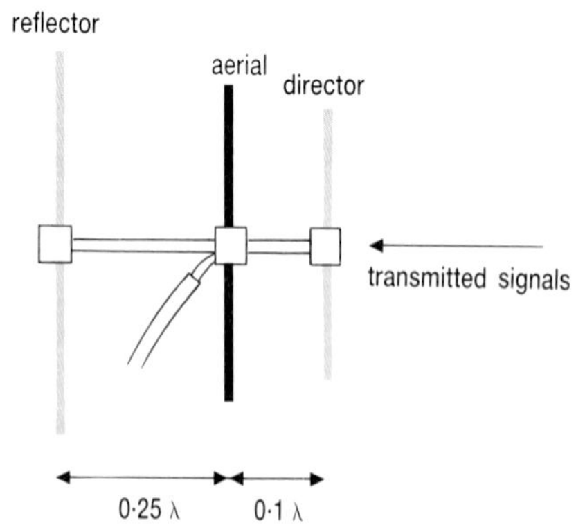

Fig 4.22 Increasing the signal strength at the receiving aerial.

Radio waves can travel through solid materials such as brick with very little attenuation. You could look at this with the transmitter and receiver in the following investigation. However metal objects can seriously affect the signal due to reflection effects. Because of this it is best to position an aerial on the outside of the building, on the side closest to the transmitter. If there is a loft in the house this is also suitable.

INVESTIGATION

Testing a dipole aerial

With the frequencies used in the VHF band, aerials are rather large for the laboratory. But it is possible to use an aerial with higher frequency waves. The frequency of the oscillator used is 1 GHz or 10^9 Hz.

You will need: two dipole aerials of 0.15 m total length and stands; 1 GHz oscillator; lengths of thin metal rod (see below); sensitive galvanometer; diode.

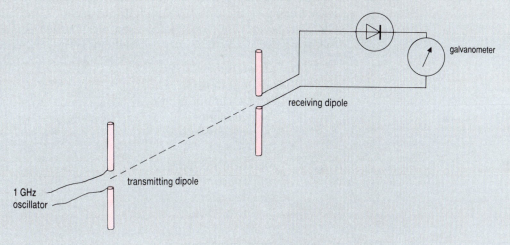

Fig 4.23   Testing a dipole aerial.

1. Set up the apparatus as illustrated in Fig 4.23. Adjust the separation until a suitable reading is obtained on the galvanometer.

2. Investigate the received signal strength in the horizontal and vertical planes of the transmitter. Is it similar to those shown in Fig 4.21?

3. Using the values in Fig 4.22, make a suitable reflector and director with the thin metal rod. Investigate the effects of bringing the reflector up behind the receiving dipole. The rods should be mounted on wooden stands and your body should be well away from the aerials when measurements are being made. At what distance behind the dipole does the reflector produce the maximum signal? Repeat this with a director. Do the distances agree with the values as calculated using Fig 4.22?

QUESTIONS

4.5 (a) Why are ferrite rod aerials not used for radio reception in motor cars?
(b) How would you use a ferrite rod aerial on a portable radio to locate the direction of a particular transmitter?
(c) How would you attempt to find out its actual position using the same arrangement?

4.4 OTHER TYPES OF RADIO TRANSMISSION

Radio users

You will see from Table 4.2 that radio broadcasting uses only a fraction of the available frequencies in each of the LF, MF and VHF bands. There are many other users in these bands and the other bands in the radio spectrum. Each frequency band is sub-divided into sections and all the sections are allocated to specific users. Some examples of the users are shown in the following list:

> armed forces
> amateur radio
> aircraft communications
> car telephones
> citizen's band radio
> fire services
> merchant shipping
> military communications
> mobile systems
> paging devices
> police
> radioastronomy
> research
> satellite communication
> weather stations

Table 4.2 Frequency allocation for radio broadcasts

Band	Frequency range	Radio broadcasts
LF	30–300 kHz	140–283 kHz
MF	300–3000 kHz	526–1606 kHz
VHF	30–300 MHz	88–108 MHz

There are about 1 000 000 licensed radio amateurs in the world. Apart from broadcasting in all the available frequency bands they even have access to a number of satellites. These are called OSCAR (orbital satellite carrying amateur radio). The latest is called OSCAR-10, being the tenth in the series.

Table 4.3 indicates how the different bands are used. (Amateur radio is not shown for the reason that sections within each of the bands listed have been allocated for their use.) Since the propagation characteristics depend on the frequency, some bands are reserved for special uses. For example the HF band is used for long distance transmissions and the VHF band for local ones. When all the available frequencies have been allocated in a particular band the next one up is considered. As demand increases and technology develops the pressure on bandwidth is ever upward. Allocations of frequencies has been made by international agreement up to 275 GHz but it will be some time before these are being used in practical systems.

Table 4.3 Uses of the radio spectrum

Band	Use
LF (3–30 kHz)	long distance communication
MF (30–300 kHz)	maritime local AM broadcasting
HF (3–30 MHz)	mobile transmissions international broadcasting maritime citizen's band radio control (models) radio telegraphy radio telephony
VHF (30–300 MHz)	local FM broadcasting TV aircraft, police, emergency services radio astronomy mobile radio telephony
UHF (300–3000 MHz)	microwave transmissions satellites TV mobile radio telephones
SHF (3–30 GHz)	microwave transmissions satellites

Television

AM radio broadcasting uses 9 kHz channels. In the MF band there is room for about 100 channels. Television requires a 6 MHz bandwidth and most TV broadcasts are placed in the UHF section of the radio spectrum to accommodate them. The channel bandwidth is 8 MHz and there are 44 channels available in the UHF band for television. In 1989 BBC and IBA broadcasts could reach 99.3 per cent of homes in the UK using 50 high-power transmitters and 750 low-power relay stations.

Microwaves

Although the first use of the microwave network was for transferring television signals around the country, nowadays it is used for telephone circuits and data transfer as well. Only about one third of its capacity is used for television and the rest carries about 20 per cent of the total signal traffic on British Telecom's trunk network.

The network consists of 200 stations and covers most of the UK. There are terminal radio stations near large cities and these are linked by repeater stations about 40 to 50 km apart situated on high ground as signals travel in straight lines (often called line-of-sight propagation) at the frequencies used. There is a range of frequencies used between 1 and 19 GHz. All the transmitting and receiving aerials are dish aerials. These are discussed further in Section 5.3. The advantages of using higher frequencies is that there is more bandwidth available and size of the aerials can be smaller. The main disadvantage is that the distance between repeaters must be smaller as attenuation rates increase with frequency. There are problems with attenuation of the signals due to fog, rain or snow. There is both an analogue and a digital network. At the moment the digital one is much smaller but eventually the complete network will be digital.

Fig 4.24 The highest dish aerial in the UK – 230 m above the ground. It was erected on the Isle of Wight for communication with the Channel Islands.

4.7 Look at a modern radio. List the differences between the MF receiver described in Section 4.2 and a modern radio.

4.8 Select one of the uses for radio indicated in Section 4.4. Try to find out which frequencies are used and why, the number of channels employed and what type of equipment is needed.

SUMMARY

The reception of radio waves first requires selection of the desired frequencies and this is commonly done using a parallel inductor–capacitor circuit. In the simplest receiver the AM transmissions are demodulated before being fed to an earpiece. Improvements can be made by changing the aerial design and adding amplification. Transmitting and receiving aerials are very similar in principle. In the LF and MF bands metal or ferrite rod receiving aerials are used, while dipole aerials are used in the VHF band. Aerial gain can be increased by adding extra metal elements. Apart from the broadcasting bands the radio spectrum contains a large number of other users.

QUESTIONS

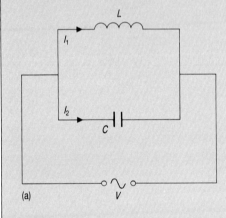

(a)

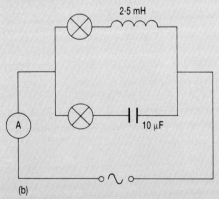

(b)

Fig 4.25

4.9 **(a)** Fig 4.25 (a) shows an inductor of inductance L and a capacitor of capacitance C connected to a supply which provides a siusoidal alternating p.d. of constant r.m.s. value V at a frequency that can be varied. A frequency can be found at which magnitudes of I_1 and I_2 are the same. This is the resonant frequency of the circuit f_r. Write an expression for the magnitude of each current, and show that the resonant frequency is given by $f_r = 1/2\pi\sqrt{LC}$

(b) In the circuit shown in Fig 4.25(b), the values of inductance and capacitance are as shown giving rise to a resonant frequency of 1000 Hz. As the frequency is adjusted towards 1000 Hz, it is noticed that the current recorded by the a.c. ammeter falls to zero although the lamps are fully lit. When asked to comment on this observation, a student suggested that the currents in the capacitor and the inductor are in opposite directions at all frequencies, and at a frequency of 1000 Hz, these currents are equal and there is no current in the ammeter.
   Write a short article which amplifies and explains this suggestion. Your article is intended to help a student who has just started an A-level physics course.   (UCLES spec.)

4.10 In an experiment to determine the resonant frequency of a parallel inductor–capacitor circuit the following results were obtained.

frequency /kHz	p.d. /V	frequency /kHz	p.d. /V
38.00	0.90	41.00	6.40
39.00	1.40	41.25	4.53
39.50	1.90	41.50	3.33
40.00	3.03	42.00	2.07
40.25	4.13	43.00	1.13
40.50	6.07	44.00	0.73

(a) Plot a graph of p.d. against frequency and from this estimate a value for the resonant frequency, f_r.

(b) What change would you make in taking the readings to enable you to get a more accurate graph?

(c) The 'sharpness' of resonance is measured by the Q-factor, where $Q = f_r/\Delta f$ and Δf is the width of the resonance curve at

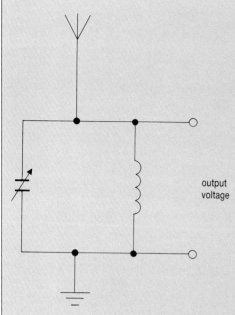

Fig 4.26

$1/\sqrt{2}$ of its maximum value. Calculate the Q-factor.

4.11 The diagram (Fig 4.26) shows the first stage of a radio receiving system.

   (a) State the function of the parallel L–C circuit. Sketch a graph to show how the output voltage varies over a wide range of signal frequencies of constant amplitude. Discuss the significance of the shape of your graph.

   (b) If the capacitor in a tuning circuit has a capacitance of 250 pF, find the inductance of the inductor when the radio receiver is tuned to BBC Radio 3 broadcasting at a frequency of 1.215 MHz.   (ULSEB spec.)

4.12 Fig 4.27 (a) shows the aerial and tuning circuit of a simple radio receiver which can be tuned to two frequency bands. With the switch in position **A** and the capacitor **C** at the middle of its range, the circuit is tuned to a frequency f.

  (a) State whether each of the following changes, on its own, would increase or decrease the value of f.

    (i) increasing the capacitance of **C**.

    (ii) switching from position **A** to position **B**.

  (b) (i) State two effects caused by the addition of resistance to a tuned circuit.

    (ii) Explain why it is more difficult to tune a domestic receiver precisely to a station on the short-wave band than to one on the medium-wave band.

  (c) Fig 4.27 (b) shows a block diagram of a simple radio receiver for amplitude midulated (AM) signals.

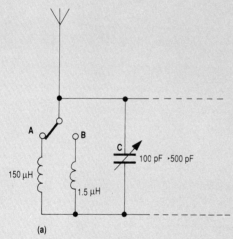

(a)

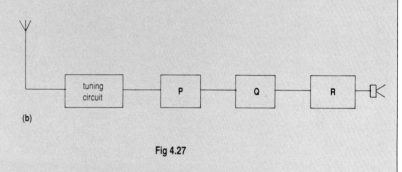

(b)

Fig 4.27

    (i) Identify the stages **P**, **Q** and **R**, and explain the function of each.

    (ii) Draw diagrams to show the waveform at the output of each of these stages.

  (d) Broadcast radio transmissions are either amplitude modulated (AM) or frequency modulated (FM).

    Describe the principal features of a frequency modulated (FM) radio wave. Refer particularly to the differences in the modulated waveforms of:

      (i) high and low frequency audio notes of equal amplitude;

      (ii) loud and soft audio notes of equal frequency.

    Explain why FM transmissions are confined to the VHF radio band.   (COSSEC AS 1989)

4.13 **(a)** Distinguish between amplitude modulation and frequency modulation.

    (b) The block diagram of an amplitude modulated radio receiver is shown in Fig 4.28(a).

      (i) Describe the function of the tuning circuit and the detector (demodulator).

(ii) Fig 4.28(b) represents the signal at Q. Draw sketch graphs to show the corresponding signals at R and S.

(iii) Draw a circuit diagram to show how an operational amplifier can be used to provide the required audio frequency amplification. Suggest suitable values for the circuit components if a gain of about 10 is required. (UCLES spec.)

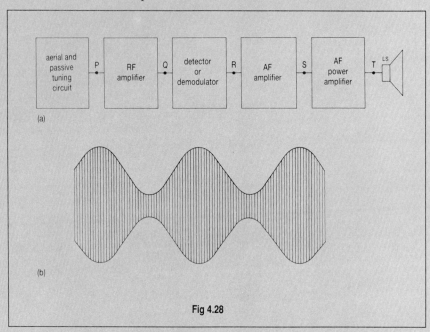

(a)

(b)

Fig 4.28

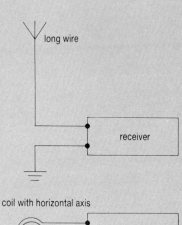

long wire

receiver

coil with horizontal axis

receiver

(a)

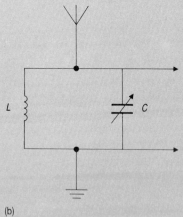

L

C

(b)

Fig 4.29

4.14 Suggest reasons for the following:

(a) Radio reception often fades when driving under a motorway bridge.

(b) Domestic radio receivers will not work inside an aeroplane.

(c) Reception of VHF broadcasts by car radio in a built-up area are often very variable.

4.15 A medium frequency radio receiver may have two kinds of aerial as shown in Fig 4.29(a): either a long straight wire or a coil. (The coil is actually wound on a ferrite rod and is usually enclosed within the receiver.)

(a) Reception of a particular station using the long straight aerial is best with the wire vertical. Explain how the radio waves alternating currents in the aerial, and why it matters that the wire is vertical.

(b) Reception of the same station using the coil aerial is good when the axis of the aerial is at right angles to the direction of the transmitter and poor when the axis of the coil points at the transmitter. Why?

(c) The aerial currents produce alternating currents in the tuning circuit of the receiver (see Fig 4.29(b)). Why can turning the knob which varies C tune in a particular signal? (OLE, Nuffield 1979.)

Theme **3**

THE NEWEST TELECOMMUNICATION SYSTEMS

This theme will look mainly at satellite and optical fibre systems. The first satellite was launched in 1957 and from this has evolved a tele-communication system that produced world-wide coverage in a period of 12 years. The first optical fibre was manufactured in 1970. The use of electromagnetic radiation in the infrared region has developed to such an extent that it is possible to send nearly 8000 telephone calls along an optical fibre less than the thickness of a human hair and the first transatlantic optical fibre link started operation in 1988. The final chapter looks at some recent developments in telecommunications.

PREREQUISITES

Before you study this theme you should have some familiarity with the following:
- Newton's Law of Gravitation.
- The refraction of light.

The control centre for the new London digital network.

The flight deck of an aircraft.

Using modern telecommunication systems.

The operations room of the European Space Agency.

Chapter 5

SATELLITE SYSTEMS

LEARNING OBJECTIVES

After studying this chapter you should be able to:

1. appreciate the difference between passive and active satellites;

2. derive expressions for the radius and orbital period of a satellite;

3. explain the importance of geostationary satellites in telecommunications;

4. recall the principles behind the establishment of a satellite telecommunication link;

5. know the types of aerial used and perform calculations involving gain and beamwidth;

6. recall the satellites that can be accessed using laboratory equipment.

5.1 THE DEVELOPMENT OF SATELLITES

When rocket technology had developed sufficiently to launch satellites their use for telecommunication purposes was soon forthcoming. The impact of satellites on telecommunication systems has been profound considering the number of years that they have been in operation. We have become accustomed to receiving live TV broadcasts from virtually all parts of the world. The Live Aid concert from Wembley London in 1985 had a potential satellite audience of 1200 million. The impact of satellites on the telephone system is perhaps not so obvious but a large percentage of our intercontinental telephone calls reach their destination via one of the satellites above us. Figure 5.1 shows the change in design of the telecommunication satellites used by INTELSAT (International Telecommunications Satellite Consortium).

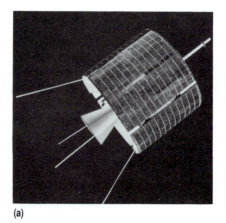
(a)

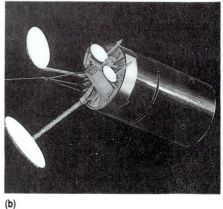
(b)

Fig 5.1   The change in INTELSAT satellites between 1965 and 1989. **(a)** INTELSAT I **(b)** INTELSAT VI. INTELSAT I would have fitted into a car boot while INTELSAT VI requires a large lorry to transport it.

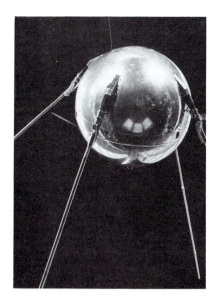

Fig 5.2 The first artificial satellite: Sputnik I, which provided the impetus for the Americans to develop their own space program leading to the first landing on the moon in 1969.

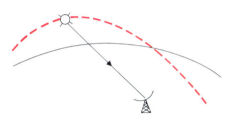

Fig 5.3 Communication with a low altitude satellite.

Passive satellites

In the years before the first satellite was launched the Moon had been used as a reflector of microwaves in research into long distance telecommunication links. Attempts were made to beam microwave transmissions at the Moon (a distance of 3.84×10^5 km away) and receive the reflected signals. The power of the returning signal was so small that the information it carried could not be easily decoded amongst all the noise it picked up on its long journey. However, the knowledge acquired proved very useful when satellite links were becoming established.

In 1957 the first successful satellite was launched by the Russians. It was called Sputnik I and carried its own transmitter for radioing information back to Earth. It sent back information for 92 days before it ceased operating. In 1960 the Americans launched Echo I. This was a plastic balloon covered with aluminium and was the first passive telecommunications satellite. A **passive satellite** does not process the transmitted signal but simply reflects it.

Active satellites

The first successful **active satellite** that could pick up the signal transmitted from Earth, amplify it and re-transmit it back was Courier 1b launched in the same year as Echo I. It was Telstar, launched two years later, that caught the public's imagination as it transmitted live television pictures between the USA and Europe. Telstar orbited the Earth at a fairly low altitude (compared with later satellites). A complete orbit took 158 minutes. This meant that communication from the ground station could only be maintained with the satellite for a short period (about 20 minutes) when the satellite was high enough above the horizon for the signals to be received. The first aerial at Goonhilly Down in Cornwall was constructed in order to work with Telstar. Not only did the aerial have to home in on the transmitted signals, it had to follow accurately the position of the satellite as it moved across the sky. This was no mean achievement considering the weight of the complete aerial assembly was over 10^6 kg.

The orbit of a satellite

The time required for one complete orbit by a satellite is called its **orbital period**. The value depends on its height above the surface of the Earth. To explain this it will be assumed that the satellite moves in a circular orbit and the Earth is a perfect sphere of uniform density. The motion of a satellite requires a centripetal force to keep it moving in a circle and that force is provided by the gravitational force of attraction between the Earth and the satellite.

The centripetal force F is given by the expression

$$F = \frac{mv^2}{R} \text{ where } \begin{array}{l} m = \text{mass of the satellite} \\ v = \text{speed of the satellite} \\ R = \text{radius of the orbit} \end{array}$$

The gravitational force of attraction F is given by

$$F = \frac{GMm}{R^2} \text{ where } \begin{array}{l} G = \text{gravitational constant} \\ M = \text{mass of the Earth} \\ m = \text{mass of the satellite} \\ R = \text{radius of the orbit} \end{array}$$

Since these forces are equal

$$\frac{mv^2}{R} = \frac{GMm}{R^2}$$

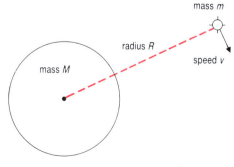

mass *m*

radius *R*

speed *v*

mass *M*

Fig 5.4 The orbit of a satellite.

From this expression, the speed of the satellite v is given by

$$v = \sqrt{\frac{GM}{R}}$$

The speed of the satellite is also given by

$$v = \frac{2\pi R}{T} \text{ where } R = \text{radius of the orbit}$$
$$T = \text{orbital period}$$

By substitution

$$T = 2\pi \sqrt{\frac{R^3}{GM}}$$

From this equation it can be seen that as the height of the satellite increases so does its orbital period.

With Telstar, the orbital period was 158 minutes. Assuming the orbit was perfectly circular its height and speed can be calculated.

Since $\quad R^3 = \frac{GMT^2}{4\pi^2}$

and given that $\quad M = 6.0 \times 10^{24} \text{ kg}$
$\quad\quad\quad\quad\quad\quad G = 6.7 \times 10^{-11} \text{ N m}^2 \text{ kg}^{-2}$
$\quad\quad\quad\quad\quad\quad T = 158 \times 60 \text{ s}$

$$R^3 = 9.16 \times 10^{20} \text{ and } R = 9.7 \times 10^6 \text{ m}$$

Since the radius of the Earth can be taken as 6.37×10^6 m the height = 3.3×10^6 m or 3300 km.

From $\quad v = \frac{2\pi R}{T}$ the speed of the satellite is $6.4 \times 10^3 \text{ m s}^{-1}$.

In practice these answers are not quite correct as the actual orbit of Telstar was slightly elliptical and the Earth is neither a perfect sphere nor of uniform density.

An interesting use of low altitude satellites is that of 'orbiting post office'. The message is sent, perhaps by explorers in a remote location, to the satellite as it passes overhead. The satellite stores the message until it passes over the intended destination where it is then transmitted down to the receiving station.

Geostationary orbits

As long ago as 1945 the scientist and science fiction writer Arthur C. Clarke had suggested that, if a satellite were placed above the equator at a height such that its orbital period was equal to the rotational period of the Earth, it would appear stationary from a point on the Earth's surface. This characteristic would enable the satellite to provide permanent coverage of a given area. Also it would simplify earth station design since it would no longer be required to track satellites moving at high speed across the sky.

By substituting the value of T as 8.64×10^4 s (i.e. 24 hours) the value of R is found to be 4.23×10^7 m. Taking the Earth's radius as 6.37×10^6 m, the height of the orbit is 3.59×10^7 m or 3.6×10^4 km. This is called a **geostationary orbit**.

Clarke had further suggested that if three such satellites were equally spaced in geostationary positions above the equator then communication coverage of most of the world would be possible except for the polar regions.

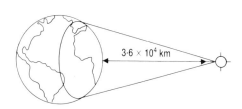

Fig 5.5 A geostationary satellite.

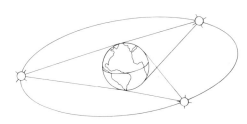

Fig 5.6 World-wide coverage with geostationary satellites.

The development of geostationary satellites

The first satellite to be successfully launched into a geostationary orbit was SYNCOM II in 1963. In 1964 INTELSAT was formed. It is owned by its member countries and is a not-for-profit commercial cooperative, providing transmission of telephony, data and TV services on a worldwide

SATELLITE SYSTEMS

basis. Inherent in INTELSAT's charter is the aim of 'contributing to world peace and understanding'. Originally formed with eleven member countries its membership in 1989 was 114. Its first satellite was Early Bird (INTELSAT I). This was launched in 1965 into an equatorial geostationary orbit above the Atlantic Ocean. The bandwidth was such that it could support 240 telephone circuits or a single television channel. Only one earth station could transmit to the satellite at a time. This is known as single access but all subsequent INTELSAT satellites have had multiple access capability. INTELSAT I had a design life of 18 months but exceeded this by a considerable margin.

In 1967 INTELSAT II satellites (the II refers to the fact that these were the second generation of INTELSAT satellites) were placed over the Atlantic and Indian Ocean regions and, in 1968, INTELSAT III satellites were positioned over the Atlantic, Pacific and Indian Oceans providing world-wide coverage. The latest generation of satellites is the INTELSAT VI series. These can support approximately 30 000 telephone circuits with 3 television channels and the first one was launched in 1989. It is important to realise that a satellite does not have an absolute capacity in terms of the number of telephone channels. The actual capacity depends on how the satellite is accessed, what modulation method is employed and how the voice signals are encoded. The INTELSAT VI satellites have an expected operational life of 10–14 years.

QUESTIONS

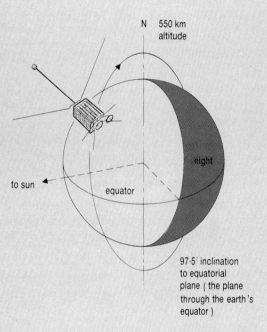

Fig 5.7 The UoSAT orbit.

5.1 Fig 5.7 shows the orbit of the UoSAT satellite (there are more details on this satellite in Section 5.4). It is a low altitude satellite and its signals can be picked up by a ground station when the satellite is passing overhead. Due to the Earth's rotation, each orbit crosses the equator 24° to the west of the previous orbit. It is called a polar orbit as the satellite passes over the north and south polar areas.

(a) From the information given, which way is the Earth rotating?

(b) If UoSAT passed overhead, how many orbits would it make before it passed overhead in the same direction again?

(c) At a single earth station there are two optimum 'windows' (periods of time when the satellite signals can be detected) each day. Explain why this is so.

5.2 ESTABLISHING A SATELLITE TELECOMMUNICATION LINK

The use of a satellite system makes very good economic sense if the cost of establishing a cable network is prohibitively high. The first such national system was established in the mid-1970s in Indonesia. This nation consists of 3677 islands of which about 1800 are inhabited. The system is called PALAPA and was ex-tended to cover the Philippines a few years later.

Launching a satellite

Satellites can be launched into geostationary orbit either by expendable conventional rocket launchers, such as the ATLAS-CENTAUR, or ARIANE, or by the manned and re-usable space shuttle. Use of either method is extremely expensive and represents approximately 50 per cent of the cost of establishing the satellite link. For the next few years most of the world's satellites will be launched by expendable rockets. The use of a space shuttle offers certain advantages as it is a re-usable vehicle with a large compartment capable of carrying up to four satellites on the same flight.

Launching the satellite by rocket is normally in three stages. The satellite is first launched into a low altitude circular parking orbit at a height of about 200 km. This will be the **perigee** (the nearest point of approach to the Earth) in the next stage. The second stage consists of changing this into a highly elliptical transfer orbit with an **apogee** (the furthest point from the Earth) at 3.6×10^4 km. The final circular orbit is produced by igniting the rocket motor on the satellite when it is passing through the apogee of the transfer orbit. At the same time the plane of the orbit is changed to the equatorial plane. If the launch is by rocket then the placing of the satellite in the low altitude orbit and subsequent conversion into the transfer orbit is achieved by the launching rocket. If the launch is by a space shuttle only the placing of the satellite into the parking orbit is achieved and an additional rocket motor system on the satellite is required to place it in the transfer orbit.

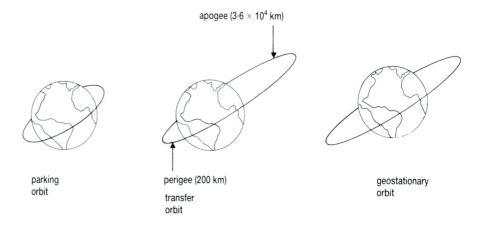

Fig 5.8 Getting the satellite into geostationary orbit.

Attitude control

It is obviously important to maintain the satellite aerials pointing at the intended areas on the Earth. This is called the **attitude** of the satellite. There are a number of ways in which the attitude can be controlled. One method (spin stabilisation) is to spin the body of the satellite about a fixed axis and to despin the aerials. The spinning produces a gyroscopic effect that maintains the spin axis parallel to the axis of the Earth's rotation. With INTELSAT VI (see Fig 5.1) the cylindrical drum on which the solar cells are mounted rotates at 50 revolutions per minute while the platform on which the aerials are mounted remains pointed at the Earth. Another method is called three-axis stabilisation. In this method the spacecraft body is in the form of a box with solar cells on panels, the aerials being mounted on three sides of the box. There are three electrically driven wheels that exchange angular momentum with the satellite so as to control its orientation. Each wheel acts independently along one of the three possible axes of rotation. If a wheel rotates in one direction then the satellite rotates in the opposite direction to conserve angular momentum.

Solar cells

A satellite can only carry a limited amount of fuel and power is required continually for operation of its electrical and electronic systems. This is supplied by the **solar cells** such as those on the drum exterior of INTELSAT VI. Solar energy falling on them is changed directly into electrical energy via photovoltaic cells made from semiconductor materials which produce a potential difference when electromagnetic radiation of suitable frequency falls on them. The efficiency of this energy conversion is initially about 10 per cent. The effect of the electromagnetic radiation on the cells results in a 30 per cent drop in this efficiency over a period of five years.

Each cell has only a small electrical output (of the order of 50 mW) and so a large number are joined together to form what is called an array. Much of the electrical energy produced is used to run the on-going functions of the satellite but the rest is used to charge the batteries that provide stand-by power for use when the satellite is in the Earth's shadow. The maximum power output of INTELSAT VI is about 2.2 kW (roughly the power consumption of a 2-bar electric fire).

Spacing the geostationary satellites

Geostationary satellites have to be in an equatorial orbit at a height of 3.6×10^4 km above the Earth's equator. Consequently there is a limit to the number of satellites that can actually be in this orbit. There is a minimum angular separation (the angle is measured from the centre of the Earth) of 3° at the moment but this will be reduced to 2° in the future. These minimum values are due to diffraction effects and are explained further in the next section. When a satellite reaches the end of its useful life the thrusters are ignited to give it a final 'kick' that moves it out of its equatorial orbit and it is then switched off by the command station.

Earth stations

Earth stations transmit and receive signals using the same aerial, often called a dish aerial. At its simplest, one earth station receives the signals from the terrestrial network, amplifies and transmits them at the uplink frequency. They reach the satellite where they are amplified, changed to the downlink frequency, further amplified and then transmitted to another earth station. The receiving earth station detects the signals, amplifies them and relays them to the terrestrial network. In the UK there are a number of earth stations used for telecommunications.

The longest established ones are at Goonhilly in Cornwall and at Madley

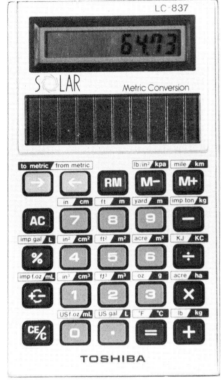

Fig 5.9 These solar cells on the calculator work in the same way as those on a satellite.

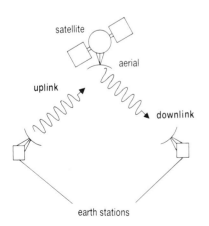

Fig 5.10 A simple satellite link.

in Herefordshire (one of the aerials at Madley is shown on the cover). In addition there is London Teleport in Docklands, which is used for television and data transmissions, and Aberdeen Teleport, which is used for the transmission of voice and data to the North Sea oil rigs. Both London and Aberdeen Teleports are known as community terminals, serving a number of different users. There are many sole user terminals which are used for office-to-office or factory-to-factory communication.

Positioning the aerial

The word aerial or dish aerial will be used throughout this chapter though the word antenna is also commonly used (as with radio). This section deals only with geostationary satellites. If the satellite were stationary relative to the Earth there would be no need to alter the position of the earth station aerial having first gained access to the satellite. In practice, because of perturbations of the Earth's orbit and gravitational effects of the Sun, the satellite does not appear stationary. If the earth station and satellite aerials are not lined up exactly, considerable loss in received signal power is suffered. Tracking of the satellite is achieved by the earth station receiving fixed frequency beacon signals from the satellite. A computer system decides whether the beacon signal is a maximum or not by moving the aerial in small increments and comparing the signal level with that before it was moved. The process is assisted by the computer having a stored model of the satellite's variations obtained from previous tracking and it then uses this information to decide which way the aerial should point in order to receive the maximum signal.

Satellites and telecommunications

With telecommunication geostationary satellites the delay between sending and receiving a message is normally of the order of 0.5 s. Because of this only one satellite link is normally allowed in any long distance telephone conversation. For instance in telephoning the other side of the world there would be only one satellite link with the remainder being made up via cable (speed of the signal along the cable is of the order of 2×10^8 m s^{-1}).

The path taken by a telephone call using a satellite link is illustrated in Fig 5.11. The call is routed from the local exchange to the international exchange. This would be by underground cable or, if long distance, perhaps through the microwave network. From the international exchange it will travel via Telecom Tower and the microwave network to the earth station. From there it is beamed up to the appropriate satellite. The re-transmitted signal from the satellite follows a similar path in reverse.

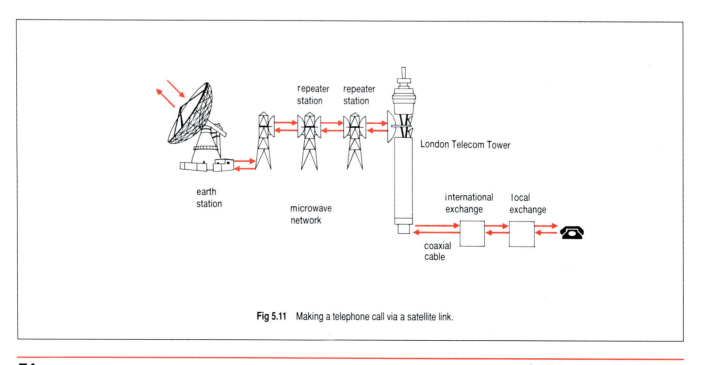

Fig 5.11 Making a telephone call via a satellite link.

SATELLITE SYSTEMS

5.3 Table 5.1 shows details of the different generations of INTELSAT satellites. Using this table describe the various developments that have been produced in satellite performance over the years.

5.4 Russian telecommunication satellites do not use equatorial geostationary orbits. They use highly elliptical orbits as much of Russia lies at very northerly latitudes.
   (a) Why is it that the satellite in an equatorial orbit is not effective at communicating with the polar regions? (Consider the angles at which the waves hit the region.)
   (b) The gravitional potential energy of a body above the Earth's surface is given by $-GMm/R$ where the symbols have the meanings already mentioned. In an elliptical orbit, assuming the total energy of the satellite remains constant, how will its speed vary as it orbits the Earth?
   (c) Using your answer to **(b)** where would the perigee and apogee of a Russian telecommunication satellite be with respect to that country?

Table 5.1   Evolution of INTELSAT satellites

	INTELSAT I	INTELSAT II	INTELSAT III	INTELSAT IV	INTELSAT V	INTELSAT VI
Year of first launch	1965	1967	1968	1971	1980	1989
Width dimension m (undeployed)	0.7	1.4	1.4	2.4	2.0	3.6
Height dimension m (undeployed)	0.6	0.7	1.0	5.3	6.4	5.3
Design lifetime, years	1.5	3	5	7	7	13
Capacity (voice circuits	240	240	1500	4000	12 000	30 000
TV channels	or 1	or 1	or 4	+ 2	+ 2	+ 3
Bandwidth MHz	50	130	300	500	2250	3300

Table 5.2   Frequency bands used with telecommunication satellites

	6/4 GHz Band	14/11 GHz Band	30/20 GHz Band
Uplink	5.925–6.425 GHz	14.0–14.5 GHz	29.5–31 GHz
Downlink	3.7–4.2 GHz	10.95–11.2 GHz 11.45–11.7 GHz	19.7–21.2 GHz
Bandwidth	500 MHz	500 MHz	1.5 GHz

5.3 TRANSMITTING INFORMATION

Frequencies used

At present there are two main frequency bands used for satellite telecommunications. The first one to be used was the 6/4 GHz band. The first figure refers to the uplink frequency and the second to the downlink frequency. You will notice from Table 5.2 that satellite telecommunications are allocated fixed bandwidths for the uplinks and downlinks. Other sections are given over by international agreement to other satellite users such as INMARSAT (see Section 7.3), military communications and for scientific purposes. There is an important reason for the uplink and downlink frequencies being different. When the signal is actually broadcast from the transmitting aerial it is sent at high power. If the same aerial or one nearby were receiving at the same frequency then the outgoing signal could swamp any low power incoming signal.

As the technology developed another 500 MHz was allocated in the 14/11 GHz band. To date most satellites have worked in the 6/4 GHz band but newer systems are working in the 14/11 GHz band as well as the 6/4 GHz. One important advantage in the 14/11 GHz band is that receiving aerials can be much smaller and thus cheaper. However, at these frequencies the signals are more susceptible to fading due to rain or snow. As technology develops the next band to be used will be in the 30/20 GHz range and there will be 1.5 GHz of bandwidth available for uplink and downlink frequencies. Although still experimental the first commercial systems are now appearing.

As the frequencies are in the same range as those used in the terrestrial microwave network the propagation characteristics are similar. The main difference between the two systems is the much greater distance between transmitter and receiver, which means far greater attenuation of the signal and much more noise is added to the signal. Also earth stations pick up a lot of noise from space. The frequencies used in the 30/20 GHz range suffer much greater attenuation than the lower frequencies due to the effects of rain or snow and absorption by molecules in the air.

Modulation and multiplexing

Although frequency modulation (FM) is most widely used at the moment, signals are increasingly being encoded digitally using pulse code modulation (PCM) and transmitted using phase shift keying (PSK) on the carrier. The type of multiplexing depends on whether the signals are analogue or digital. FM signals use frequency division multiplexing (FDM) and PCM signals use time division multiplexing (TDM). There are a number of variations in coding and modulation in present use.

Aerials

Signals sent between transmitting and receiving aerials in a satellite link suffer a great deal of attenuation because of the distance involved. The net loss in signal power with a geostationary satellite is of the order of 200 dB. This would mean that a transmitted 10 W signal from a ground station would reach the satellite with a power of 10^{-19} W. Also noise is added to the signal and for satisfactory decoding of the signal a minimum signal-to-noise (S/N) ratio must be exceeded. With digital signals it may be possible to get away with only a few errors having an S/N ratio of 10 dB. Receiving satellite TV broadcasts requires at least 20 dB.

While nothing can be done to eliminate the actual loss in signal power during transmission the design of the aerials can greatly improve the amount of detected signal power. The function of any aerial is to produce gain (as with the dipole aerial in Section 4.3). With the transmitting aerial the aim is to send the signals in a specified direction while the aim of the receiving aerial is to collect as much of the signal power as possible. Directional properties are very important in satellite communications because of the distances involved. Although dipole aerials with directors and reflectors provide sufficient gain in the VHF band (and are used with the low altitude polar orbiting satellites that broadcast in this band – see Section 5.4), they will not provide sufficient gain at the frequencies used with geostationary telecommunication satellites. The 'dish' of the dish aerial acts as a reflector of the waves. The shape of the reflector is actually

Although the bandwidth in satellite telecommunications is 500 MHz in the 6/4 and 14/11 GHz bands, the receiver/amplifier/transmitter sections (called transponders) on the satellite are designed to operate over only a range of 40 MHz. Normally there will be 12 transponders working but the satellite carries spare ones that can be switched on if a working one fails.

parabolic rather than spherical. The reason for this is illustrated in Fig 5.12. All the waves emitted from the focus of a parabolic reflector emerge parallel to each other. Contrast this with the spherical surface where only the central waves remain parallel. The effect is similar if the waves hit the reflecting surface as a parallel beam. Only with a parabolic reflector will they be focussed at a point.

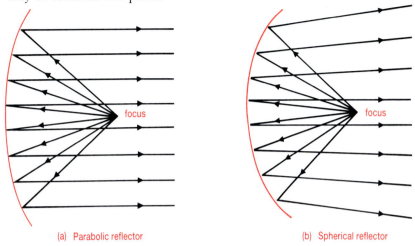

(a) Parabolic reflector   (b) Spherical reflector

Fig 5.12 Parabolic and spherical reflectors.

Although both satellite and ground station aerials are parabolic in shape they differ in size and the way of feeding the signals in and out of the aerial system. Unlike the earth station the size of satellite aerials is limited by the launching vehicle. The signal is fed to or from the dish on the satellite by means of a horn feed. With the satellite aerials the microwaves are fed through a metal waveguide (hollow metal tube) to the horn, which is placed at the focus. With the larger ground stations there is a small metal reflector placed at the focus which directs the beam to or from the waveguide. This is usually called a Cassegrain feed.

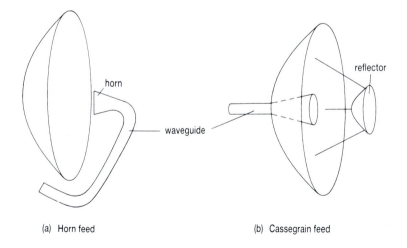

(a) Horn feed   (b) Cassegrain feed

Fig 5.13 Two types of aerial feed.

Aerial gain

The gain of a dish aerial can be expressed as the ratio of the actual signal power sent out in (or received from) a particular direction to that transmitted by an aerial radiating uniformly in all directions. This ratio is usually expressed in decibels. For instance, if the gain of a transmitting aerial is 60 dB, this means that the transmitted signal in a particular direction is increased by a factor of 10^6.

The practical gain of a parabolic reflector is given by the expression:

$$\text{Gain (in dB)} = 10 \log_{10} 6(D/\lambda)^2$$

where D = diameter of the dish
λ = wavelength of signal

For example, the gain of a 32 m diameter aerial operating at 6 GHz would be about 64 dB. In addition to the aerials there are high gain electronic amplifiers on the satellite and the earth stations. These amplifiers are essential if signals are to be satisfactorily transmitted and received as the gain provided by the aerials alone is not sufficient.

Diffraction

It has been assumed that the beam emitted from a parabolic transmitting aerial travels in a straight line. However, when waves pass through an aperture diffraction occurs (this wave effect was mentioned in Chapters 2 and 3). The dish aerial can be considered as a circular aperture and diffraction causes the waves to spread out beyond the edges of the aerial dish as illustrated in the polar diagram in Fig 5.14. (Polar diagrams were discussed in Section 4.3.)

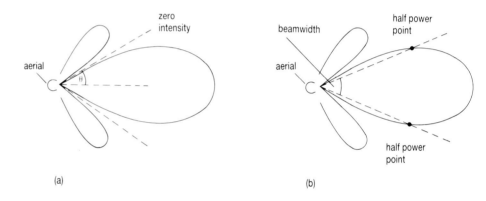

Fig 5.14   Diffraction through a circular aperture.

For a circular aperture the first minimum occurs at an angle θ given by the expression:

$$\sin \theta = \frac{1.22 \, \lambda}{D}$$

where λ and D have the same meanings as in the last equation

Compare this with the formula mentioned in the investigation in Section 3.4 for diffraction through a single slit.

In telecommunications the amount of diffraction is usually expressed using the term **beamwidth**. This is defined as the angle between the two directions at which the transmitted power (measured from the direction of greatest intensity) has dropped to 0.5. This is often called the half power or − 3 dB point and leads to the expression:

$$\text{Beamwidth (in degrees)} = \frac{70 \, \lambda}{D}$$

When signals are transmitted from an earth station the effect of diffraction means that they could be received by more than one satellite. To eliminate this there has to be a minimum angular separation between geostationary satellites in the equatorial plane. At the moment the angular separation is 3°. This means that it would be theoretically possible to have 120 satellites in orbit simultaneously.

Footprints

The area to which the satellite aerial transmits on the Earth's surface is called its **'footprint'**. It is possible, by changing the position of the horn feed or using different aerials, to vary the size and shape of the footprint. The different sizes of footprint range from 'global' (where the transmission of the signal covers the whole area visible to the satellite) to 'spot' (where the signal is directed towards a small circular area covering, say, a certain country). When a satellite aerial is directed permanently towards one area the aerial is often shaped so that the footprint approximates to the area of transmission.

The bandwidth of a satellite can be effectively doubled by transmitting signals that are polarised at right angles to each other. Each of the two receiving earth stations is designed to receive only one plane of polarisation and identical frequencies can be sent to each.

QUESTIONS

5.5 (a) Show that a 10 W signal experiencing a 200 dB loss would have a received power of 10^{-19} W.

(b) If the noise background is equivalent to 10^{-22} W and the required signal-to-noise ratio is 20 dB, can the received signal be satisfactorily decoded?

5.6 (a) Calculate the gain of a satellite aerial of diameter 3 m operating at a frequency of 11.5 GHz

(b) Calculate the beamwidth of the transmitted signal from a 12 m diameter aerial assuming the frequency of transmission was 6 GHz.

(c) In an article it was said that, if the direction of the receiving aerial at a ground station were out by 0.1°, the signal power would fall by one half. Using suitable values discuss the accuracy of this statement.

(d) Why do you think the uplink frequencies are always higher than the downlink frequencies? (Think about the relative sizes of the aerials.)

(e) Compare the relative advantages and disadvantages of a satellite and radio telecommunication system.

5.4 ACCESSIBLE SATELLITES

This section will look at some satellites that, provided you have the requisite equipment, can be accessed in your school or college. All except one are low altitude satellites and broadcast in the VHF band. Their signals can be picked up with dipole-type aerials. The signals transmitted are **circularly polarised**. Unlike the polarisation used with radio broadcasting transmissions where the plane of polarisation is fixed, the plane of polarisation appears to rotate about the direction of propagation. The

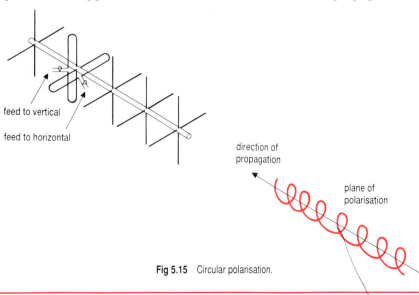

feed to vertical

feed to horizontal

direction of propagation

plane of polarisation

Fig 5.15   Circular polarisation.

reason for this type of transmission is that low altitude satellites do not have the stability of the geostationary ones. In practice this means a type of double dipole aerial is required to receive the signals.

UoSAT-1 and-2

These satellites were designed and built by the Department of Electronics and Electrical Engineering at the University of Surrey. The main experimental functions of the satellite are: propagation studies of VHF signals and microwaves, detection of electrons, investigation of the Earth's magnetic field and the production of 'Earth pictures'. Each was launched by NASA: UoSAT-1 in 1981 and UoSAT-2 in 1984 (a third satellite will be launched soon). A UoSAT satellite is illustrated in Fig 5.7. The comprehension passage at the end of this section is about UoSAT-2.

NOAA Weather satellites

The term NOAA refers to the National Oceanic and Administration Organisation in the USA. These satellites have been launched at intervals since the early 1960s. The latest is NOAA-10. All have polar orbits but their orbits are inclined at angles to each other. All transmit weather pictures in the VHF band. There are sensors on board which respond to visible or infrared radiation and the satellite produces both visible and infrared pictures (the infrared shows very hot areas as black on the received picture through a range of greys to white for very cold). The picture is transmitted at 120 lines per minute. As the satellite moves along in its orbit the transmitted picture is continually updated.

METEOSAT

METEOSAT is the only geostationary satellite that can be accessed by reasonably cheap equipment although a dish aerial is needed. It is a meteorological satellite and broadcasts in the UHF region at 1.6945 GHz. While the polar orbiting satellites are used for short term weather forecasting the geostationary ones are used for longer term predictions.

Fig 5.16   Downloading pictures from METEOSAT.

SATELLITE SYSTEMS

One of the two METEOSAT satellites is positioned above the equator on the Greenwich meridian. It transmits signals down to the Operations Centre of the European Space Agency in West Germany where the pictures are processed by computer and the outlines of countries superimposed before beaming the pictures up to the satellite again. The satellite retransmits them to all receiving ground stations in its footprint. The transmitted picture is changed every half hour.

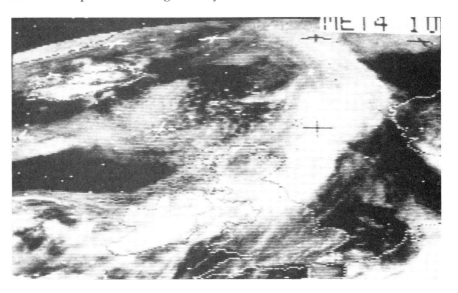

Fig 5.17 A weather picture from METEOSAT (The UK is in the centre of the picture).

COMPREHENSION

(This is taken from the booklet *UoSAT: The University of Surrey Satellite Project*.)

Telecommand from the ground is carried out by digitally-coded radio signals, but in order to prevent unauthorised persons interfering with the spacecraft the craft will only act on messages preceded by a long secret code word. This code word also prevents the satellite from misinterpreting and acting on random interference, such as that from powerful radar transmitters.

The telemetry system and the satellite's downlink provides a means of comprehensive surveillance of the on-board system for engineering purposes and for receiving information from the experiments. It also provides a channel for the camera experiment. There are several data formats possible to cater for a variety of different ground station facilities and also to maximise the reliability of communication with the spacecraft. An important consideration has been to make it possible for the telemetry to be received by relatively simple receivers and aerial systems.

UoSAT can send out data using ASCII codes at 1200, 600, 300, 110 or 75 bauds, as well as radio teletype at 45.5 bauds and 10 or 20 words per minute Morse. It can also send out a limited range of messages using an electronically synthesised voice.

The telemetry system makes measurements of 60 analogue voltages or currents, representing the outputs of the various experiments, or the details of behaviour of the component parts of the spacecraft, such as the solar array current, or the temperature of the battery pack. In addition it measures the status of 45 points on the satellite. These give data as to their switched position, on or off, and thus are digital in nature. The whole collection of measurements are made and transmitted in succession, one complete frame of data taking about 5 seconds at the fastest rate of transmission of 1200 baud. However, on command the system can dwell on one status point to examine it more closely. Most of the channels have an accuracy of ± 2%.

The primary radio links operate at VHF (145.825 MHz) and at UHF (435.025 MHz). The modulation used is narrow-band frequency-modulation (n.b.f.m.), with about ± 2 kHz deviation. This is easily received and demodulated by a standard narrow-band VHF or UHF amateur receiver, and a small fixed crossed-pole antenna (to receive the circularly polarised signal) should be adequate for most orbit passes. The addition of a 10 dB gain Yagi antenna steerable in azimuth will provide coverage for low-elevation passes.

The demodulated output from the receiver reproduces the original data signal which is sent using phase-synchronous audio-frequency shift-keying. When sent at 1200 bits s^{-1} the data comes in the form of 1200 Hz tones for '1' and 2400 Hz for '0'. The transitions occur at the zero crossings of the tone waveforms; thus exactly 2 cycles of 2400 Hz represent 0. Fig 5.18, for example, illustrates '1-0-1'.

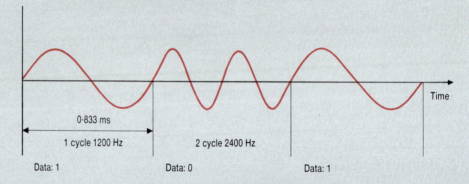

Fig 5.18   Data waveform for 1 – 0 – 1 (phase synchronous).

1. Use a telecommunications dictionary or reference books, if necessary, to help you find out the meaning of the following words and terms:
    (a) random interference
    (b) telemetry
    (c) ASCII code
    (d) bauds
    (e) narrow band receiver
    (f) azimuth

2. What analogue quantities (apart from those mentioned in the article) would it be useful to measure?

3. How many binary digits would be sent during transmission of one complete frame of data?

4. What is meant by the term 'narrow-band frequency-modulation'?

5. Describe what circular polarisation is and the type of aerial needed to receive it.

6. Explain the meaning of the term '10 dB Yagi antenna'? What would be the detected power of a received 10^{-6} W signal?

7. Draw the waveform for the data 1-1-0-1-0 using phase-synchronous shift-keying.

The first active satellite introduced a new era in telecommunication systems. The geostationary satellite allowed continuous communication between two ground stations. The large distances involved problems in receiving the transmitted signals. Accurately aligned aerial dishes with large gains are essential together with high gain amplifiers. Satellites are now an integral part of the telecommunication system. Satellites are now operating with a wide range of functions.

QUESTIONS

5.7 A 10^{-18} W signal is received by a satellite aerial of diameter 2 m. Within the satellite it is further amplified by 100 dB before being transmitted back to Earth via a similar aerial. Assume that the up and downlink frequencies are 6 and 4 GHz respectively.
Calculate: (a) the gain of each aerial,
(b) the gain of the complete satellite system,
(c) the power of the transmitted signal.

5.8 An aerial of diameter 30 m is operating at 4 GHz.
(a) Calculate: (i) the gain of the aerial,
(ii) the angular spread of the main lobe,
(iii) the beamwidth.
(b) Using your answer to (c) (iii), what would be the diameter of the beam at the height of a geostationary satellite?
(c) Assuming the separation of the geostationary satellites must be at least equal to the beamwidth of the transmitted signal, how many satellites could theoretically be in such an equatorial orbit?

5.9 'INTELSAT uses communication satellites in circular geosynchronous orbits about the equator of the Earth. An earth station transmits telephone signals as modulated carrier waves on particular carrier frequencies allocated to it in the band 5.925 GHz to 6.425 GHz, and receives from the satellite particular carrier frequencies in the band 3.700 GHz to 4.200 GHz'.
(a) What is a geosynchronous orbit?
(b) What do you understand by the term modulated carrier wave? What is the wavelength of the carrier wave if its frequency is 6.400 GHz? Into which particular region of the electromagnetic spectrum does this radiation fall? (The speed of light in vacuum $c = 3.00 \times 10^8$ m s^{-1}.)
(c) Why is amplification needed at the satellite? Why do satellites also transmit beacon signals? (ULSEB 1988)

5.10 (a) The polar diagram in Fig 5.19(a), which is not to scale, shows the 'radiation pattern' from a parabolic dish transmitting aerial centred at O with its axis along OA.
(i) Explain carefully what is represented by such a diagram and point out any significant features.
(ii) The angle is found to be 8.0° when the dish is transmitting a signal of wavelength 0.23 m. What is the diameter of the dish aerial? You may assume the 1.22 factor for circular diffraction minima.
(b) (i) An electromagnetic wave of frequency 10.00 GHz travelling in free space enters a medium of relative permittivity 1.610 at this frequency. Use the data below to determine
1. the speed of propagation in free space
2. the wavelength of the wave in free space. Express your

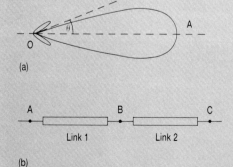

(a)

(b)

Fig 5.19

answers to four significant figures.
(Permeability of free space = $4\pi \times 10^{-7}$ H m^{-1} permittivity of free space = 8.8542 $\times 10^{-12}$ F m^{-1}).
State how the speed of propagation changes when the wave enters the medium from free space.

(ii) At a certain point in free space an electromagnetic wave is propagating as a plane wave. Give a labelled sketch to illustrate the E and B fields of the wave.

(c) What do you understand by the terms
(i) *noise*, and
(ii) *signal-to-noise* ratio in a communication system?

(d) A communication system consists of two links between a source at A and a receiver at C as shown in Fig 5.19(b). In link 1 and link 2 there is no loss of signal power, but on joining link 2 at B half of the total input power at that point is lost. In each of the links, noise is generated with a signal-to-noise ratio of 10^5. If a noise-free signal of power P enters link 1 at A, what is
(i) the input power entering link 2,
(ii) the signal-to-noise ratio at C?   (ULSEB 1989)

5.11 (a) Long distance intercontinental telecommunications may use free-space electromagnetic wave propagation linking ground station to ground station. Use a diagram to show how this is achieved for (i) microwave signals, and (ii) radio signals. (Apparatus details are NOT required.)

(b) Estimate the transit times for a UK-USA telephone signal link using (i) satellite communication, and (ii) surface wave propagation. Assume that the satellite is 36 000 km equidistant from each ground station and that the great circle distance between the two ground stations is 8400 km. (Speed of electromagnetic waves in free space = 3.00 $\times$ 10^8 ms^{-1}.)   (ULSEB 1989)

Chapter 6

OPTICAL FIBRE SYSTEMS

LEARNING OBJECTIVES

After studying this chapter you should be able to:

1. understand the importance of low loss glass and the laser in the development of optical fibre transmission links;

2. describe the propagation of light waves in glass fibres and the problems associated with this propagation;

3. recall the advantages and disadvantages of the three main types of optical fibre;

4. describe the standard optical sources and detectors used in optical fibre links;

5. explain the functions of the sub-systems of a complete optical fibre system;

6. recall the advantages and disadvantages of optical fibre and coaxial systems.

6.1 THE DEVELOPMENT OF OPTICAL FIBRE SYSTEMS

Some of the original pre-electrical methods of telecommunication used light as the carrier. The flashing of sunlight from polished metallic surfaces had long been an accepted method of conveying simple information. For a long time lighting fire beacons was one method of sending information. The approach of the Spanish Armada in 1588 was signalled across England in this way. The Aldis lamp was used until very recently on naval ships for communicating. There were limitations on using light as the carrier and its use decreased with the coming of electrical methods of telecommunication. However with the present generation of telecommunication systems, it is the frequency of the light used that is the important factor as it leads to much higher rates of information transfer.

In 1880 Alexander Graham Bell demonstrated a device for transmitting speech via the modulation of a light beam. He called it a photophone and a transmission distance through air of over 200 m was achieved.

As we have seen with radio and satellite systems the move is always towards utilising higher and higher frequencies. Comparing radio waves at 30 GHz (3×10^{10} Hz) with the frequency of red light (4×10^{14} Hz), the frequency increase is of the order of 10^4 times. Hence the possibility of greatly increased bandwidths for information transfer. Strictly speaking, unless a type of modulation such as phase shift keying is used, the full capability of the available bandwidth cannnot be used. In present systems the technology only allows amplitude modulation, which does not fully exploit this bandwidth. Even transmitting nearly 8000 telephone calls along a fibre less than the thickness of a human hair is not fully exploiting the available bandwidth.

Glass as a transmission medium

There were two main reasons why light had not been used until recently for telecommunication purposes. Firstly there was no suitable medium

Fig 6.1 The signal capacities of the optical fibres on the left and the copper cables are the same, but their size is not!

available for long distance transmission. The problems associated with sending light through the air to a receiver are very apparent. Light waves can be considered to travel in straight lines and so the transmitter and receiver must be in view of each other. Also light waves are absorbed by fog, rain or other adverse weather conditions. In 1966 Kao and Hockham, working at the company Standard Telecommunication Laboratories, demonstrated that glass fibres would make a suitable medium for transmission over long distances provided that the glass could be made pure enough to reduce the light attenuation to an acceptably low level. Glass fibres have become the accepted method for light communication systems. Secondly no suitable light source was available.

Glass fibres

Glass may appear transparent to us because we normally only look through thin sheets of it. Only when you look through a sheet of glass lengthways do you notice its lack of transparency. The reason for this lack of transparency is the presence of impurities in the glass. The transmitted light may sometimes look greenish due to the presence of copper and iron in the glass. Normal glass will attenuate light waves at a rate of 100 dB km^{-1}. This means that light passing through a 1 km thickness will be reduced in power by 100 dB – a reduction by a factor of 10^{10}. Kao and Hockham said that an attenuation of 20 dB km^{-1} would enable a suitable long distance system to be set up. By removing most of the impurities this level was reached in 1970. Now 0.2 dB km^{-1} is possible. With optical fibres it has become accepted practice to use wavelength rather than frequency and this convention will be used. This is initially rather confusing as frequency was the accepted way when dealing with radio waves.

Power loss as an exponential change

The power loss in a glass fibre can be expressed in another form. As the light waves progress through a medium a constant fraction of the light power is lost per unit distance travelled. For instance 5 per cent may be lost in travelling through 0.1 m. In the next 0.1 m 5 per cent of the then existing power is lost and so on. This is an example of an exponential decrease and you have probably met these in other areas of physics such as radioactive decay or the discharge of a capacitor through a resistor. This exponential decrease may be represented by the following equation:

$$P = P_0 e^{-\alpha x} \text{ where } P_0 = \text{power at } x = 0$$

$$P = \text{power at distance } x$$

$$\alpha = \text{attenuation coefficient}$$

By taking natural logarithms the expression becomes:

$$\ln P = \ln P_0 - \alpha x$$

If $\alpha = 0.1$ km^{-1} then the power of a 1 mW signal after it has travelled 2 km is equal to:

$$\ln P = \ln 10^{-3} - 0.1 \times 2$$

$$\ln P = -6.91 - 0.2 = -7.11$$

So $P = 8.2 \times 10^{-4}$ W or 0.82 mW

$P = P_0 e^{-\alpha x}$ is related to the usual expression for the power loss in decibels:

$$\text{Number of dB} = 10 \log_{10}(P_0/P) = 4.34 \ln(P_0/P)$$
$$= -4.34 \times \alpha$$

The light source

To use light for high rates of information transfer there are three main requirements for the light source. The light should be produced with a single wavelength. If a range of wavelengths are present each wavelength travels at the same speed in a vacuum but travels in glass at a slightly different speed. This produces problems that are discussed in the next section. Secondly the waves emitted should be coherent. **Coherent** means that the waves produced by the source must be in phase with each other. Light is emitted in small packets of energy called **photons**. These photons need to be in phase. The emission of radiation from the conventional light source is completely random. If there are no phase relationships between the emitted photons – the light is said to be **incoherent.** This is illustrated in Fig 6.2. This is why a conventional light source cannot be used except for low rates of signal transfer over short distances. Finally the source must be capable of being turned on and off very rapidly.

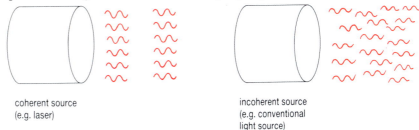

coherent source
(e.g. laser)

incoherent source
(e.g. conventional
light source)

Fig 6.2 Coherent and incoherent light sources.

The best solution to the problem of a light source was the **laser**. LASER stands for Light Amplification by the Stimulated Emission of Radiation. The explanation of how the laser works is complicated, suffice it to say that it produces light waves of virtually a single wavelength that are emitted in phase and it can be pulsed on and off very rapidly. Lasers made from semiconductor materials have been produced that emit their single wavelength (strictly speaking it is a very narrow band of wavelengths) in the visible region of the spectrum or just outside it in the infrared. For reasons that will be explained later the lasers used in optical fibre telecommunication systems work in the infrared. Usually the word 'light' would not be used for waves in this section of the electromagnetic spectrum. But common usage has decided that, provided the infrared is generated and transmitted by similar methods to those used for visible light, the infrared can be called light.

The first public telecommunication link involving optical fibres started operation in the USA in 1977 and, later that year, the first link in Europe was opened between Martlesham, which is the research centre for British Telecom, and Ipswich – a distance of 13 km. It could carry 1920 telephone calls. As most optical fibre sytems are digital in operation it is useful to know the number of bits per second that can be transmitted. The Martlesham–Ipswich link could carry 140 Mbit s^{-1} which means that 1.4×10^8 1s and 0s could be sent each second.

QUESTIONS

6.1 (a) Show that the power of a light signal would be attenuated to 1/100 of its original value after travelling 1 km through glass of attenuation rate 20 dB km^{-1}. What value would this attenuation rate give for the attenuation coefficient?

(b) What would be the power of a 10 mW light signal after it had travelled 1 km in ordinary glass (attenuation rate 100 dB km^{-1})?

(c) How many times longer would an optical fibre (attenuation rate 1.0 dB km^{-1}) have to be in order to produce the same attenuation as ordinary glass (attenuation rate 100 dB km^{-1})?

6.2 OPTICAL FIBRES

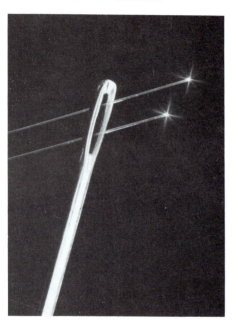

Fig 6.3 Many optical fibres could pass through the eye of a needle.

Producing the fibres

One type of glass used in the construction of optical fibres is made from silicon (II) oxide which is usually called silica and is the main constituent of common sand. When normal glass is made it contains a number of impurities and this is the reason for its lack of transparency. One common impurity is the OH$^-$ radical from water and ions of various metals such as copper, iron and chromium may be present. One method of producing fibres consists of cooling silica vapour on the inside of a silica tube and this eliminates most of the metal ions. The silica tube is subsequently collapsed at a high temperature to form a solid glass rod and then the rod is drawn out to produce the fibre The first type of fibre produced consisted of a central **core** surrounded by a second layer with a slightly lower refractive index. This second layer was called the **cladding**. The two different refractive indices were obtained by changing the amounts of various substances in the silica vapour before deposition. This process is called doping and is very similar to that used in the production of integrated circuits.

Attenuation in optical fibres

Attenuation of the signal is due to three main factors. The elimination of as many of the impurities as possible in the glass reduces attenuation in the fibre though the actual amount of attenuation varies with the wavelength transmitted. Glass is non-crystalline and inhomogeneities reflect and scatter the light. This is called **Rayleigh scattering** after its discoverer and is proportional to $1/\lambda^4$, where λ is the wavelength of the incident radiation. To minimise this effect the wavelength used should be as large as possible. Thirdly absorption by the glass itself increases with wavelength. The graph in Fig 6.4 shows the total attenuation due to these three factors. Silica has two minimum attenuation regions – about 1.3 and 1.55 µm. The left hand

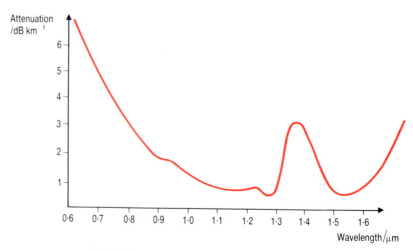

Fig 6.4 The variation of attenuation with wavelength in silica.

portion of the graph is due mainly to Rayleigh scattering and the right hand portion to absorption by the silica. The central maximum is due to absorption by the OH⁻ radicals which are virtually impossible to eliminate completely.

Another type of attenuation occurs because of irregularities between the core and cladding of the fibre produced during manufacture. Loss of power due to scattering can also occur in long distance links at the points where the sections of optical fibre are joined together. These are called coupling losses and are mainly due to slight misalignments between the joined fibres. The curvature of the cable can also produce losses if the curvature exceeds a certain value. Some losses are fixed and some are frequency dependent. With the different types of glass and plastic now used for optical fibre the transmission wavelength must obviously be selected taking into account all the frequency-dependent attenuation processes.

Total internal reflection

When light passes from one medium to another its speed changes and it is refracted. If light waves pass from a medium of lower refractive index to one of higher refractive index, they decrease in speed and are refracted towards the normal as shown in Fig 6.5(a). If there is no subscript to the left of the symbol for refractive index n then the value of n refers to light travelling from a vacuum (or air) into that medium. When light waves pass from a medium of higher to lower refractive index they are refracted away from the normal (Fig 6.5(b)).

It can be shown, for light passing from one medium to another, that:

$$n_1 \sin i_1 = n_2 \sin i_2$$

where n_1 and n_2 are the respective refractive indices i_1 and i_2 the respective angles.

For light passing from a medium of higher refractive index to one of lower refractive index, there is a certain value of the angle of incidence for which the angle of refraction is 90°. This angle is called the critical angle i_c and it can be calculated as follows:

$$n_1 \sin i_c = n_2 \sin 90°$$

As $\sin 90° = 1$, $\sin i_c = n_2/n_1$.

Any ray of light that strikes the interface at an angle greater than this critical angle will undergo **total internal reflection**. The interface behaves as a perfect mirror reflecting back rays at the same angle. At its simplest the light is continually totally internally reflected from side to side as it travels through the fibre. Any rays that strike the interface at less than the critical angle are refracted into the cladding and effectively lost.

Note that the speeds of light in different media are related by the formula:

$$\frac{v_1}{v_2} = \frac{n_2}{n_1}$$

where v_1 is the speed in the medium of refractive index n_1 and v_2 is the speed in the medium of refractive index n_2, see also Fig 6.7.

Multipath and material dispersion

The propagation of light in a fibre would appear reasonably straight forward but in practice there are complications. There is a range of angles (greater than i_c but less than 90°) at which the rays that strike the interface

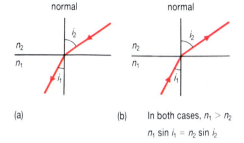

(a) (b) In both cases, $n_1 > n_2$

$n_1 \sin i_1 = n_2 \sin i_2$

Fig 6.5 The refraction of light.

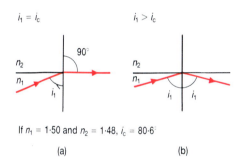

$i_1 = i_c$ $i_1 > i_c$

If $n_1 = 1.50$ and $n_2 = 1.48$, $i_c = 80.6°$

(a)   (b)

Fig 6.6 Total internal reflection.

The first practical application of total internal reflection was demonstrated by John Tyndall in 1870 at the Royal Institution in London. He showed that it was possible to transmit light along a hollow pipe filled with water.

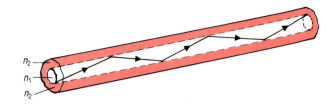

Fig 6.7 The propagation of light in a fibre.

will be totally internally reflected. Rays travelling from optical source to detector can follow a range of paths or modes as they are often called. Therefore rays from a particular pulse will not arrive at the receiver at the same time because of the differing distances travelled (this is illustrated in Fig 6.8(a) with three rays). The effect is called **multipath** (or modal) **dispersion**.

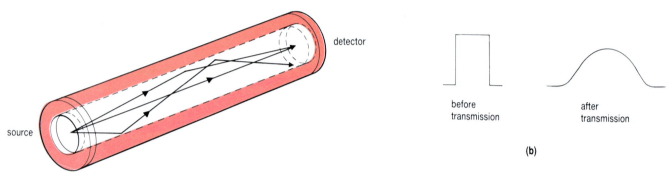

before
transmission

after
transmission

(b)

(a)

Fig 6.8 The effects **(a)** multipath dispersion and **(b)** material dispersion.

Electromagnetic waves of different wavelengths travel at different speeds in any medium apart from a vacuum. This explains why light passing through a prism is split up (dispersed) into seven different colours. Each colour has a small range of wavelengths and each wavelength travels at a slightly different speed in the glass. If the light source emits a range of wavelengths, then the waves travel at different speeds through the fibre and cause the pulses to spread out as they travel along the fibre. This is called **material dispersion**. Both these factors will produce an upper limit to the number of pulses transmitted per second and the length of fibre that can be used before regeneration of the pulses is necessary.

Because of these two types of dispersion it is often more useful to specify the **capacity** of an optical fibre in the form: bit rate × distance and the units are usually Mbit s^{-1} km. This means that, if a fibre has a capacity of 100 Mbit s^{-1} km, 100 Mbit s^{-1} can be transmitted along a 1 km length, 20 Mbit s^{-1} along a 5 km length and so on.

Types of optical fibre

There are three main types of optical fibre. Each consists of a thin central core surrounded by a layer of cladding of lower refractive index. Around the cladding there are usually two layers of plastic coating that strengthen and protect the fibre. Fig 6.9 indicates the typical dimensions of optical fibres. Most of the existing telecommunication systems use glass for the fibre but plastic fibres, which have much higher attenuation rates but are very much cheaper, are used for short distance links. Plastic fibres are usually operated in the red region (about 630 nm) of the visible spectrum where minimum absorption in the plastic occurs.

OPTICAL FIBRE SYSTEMS

Step-index fibre The step-index fibre was the first type to be manufactured. Such fibres suffer from multiple path dispersion but, because of their relative cheapness, are useful for links where the required transmission capacity is low.

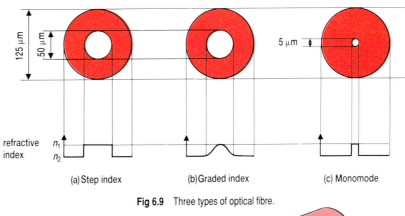

Fig 6.9 Three types of optical fibre.

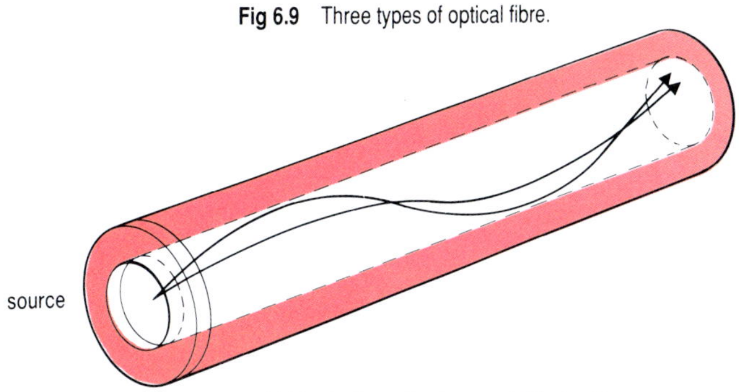

Fig 6.10 Multipath dispersion in a graded index fibre.

Graded-index fibre In this type the refractive index varies continuously across the cross-section of the core as shown in Fig 6.9(b). This helps to minimise multipath dispersion. Although the off-axis rays travel a greater distance the rays actually travel faster as the refractive index is smaller. The net effect is to reduce the multipath dispersion. This is illustrated in Fig 6.10. Graded-index fibres are more difficult to produce and consequently more expensive. Such fibres were extensively used in the early telecommunication systems but have been replaced in the more recent ones by monomode fibre. However they are still used in systems such as those mentioned for the step-index.

Monomode fibre As the name suggests there is only one transmission path (straight through i.e. axial) because of the narrowness of the core. The diameter of 5 μm is the same order of magnitude as the wavelength of the light being used. Compare this with the diameters of the 'multimode' fibres already mentioned. The narrowness of the core causes problems with getting sufficient signal power into it. This is the most expensive type of fibre and is used where high transfer rates and/or long distances are involved.

QUESTIONS

6.3 Refer to Fig 6.11 when answering this question.
   (a) Calculate the value of the critical angle for the fibre shown.
   (b) Calculate the speed of light in the core. The speed of light in a vacuum is 3.00×10^8 m s^{-1} compared with the axial ray.
   (c) Calculate the extra distance travelled by the ray hitting the interface at an angle of 81° compared with the axial ray.
   (d) What would be the difference in arrival times at point X for the two rays shown?

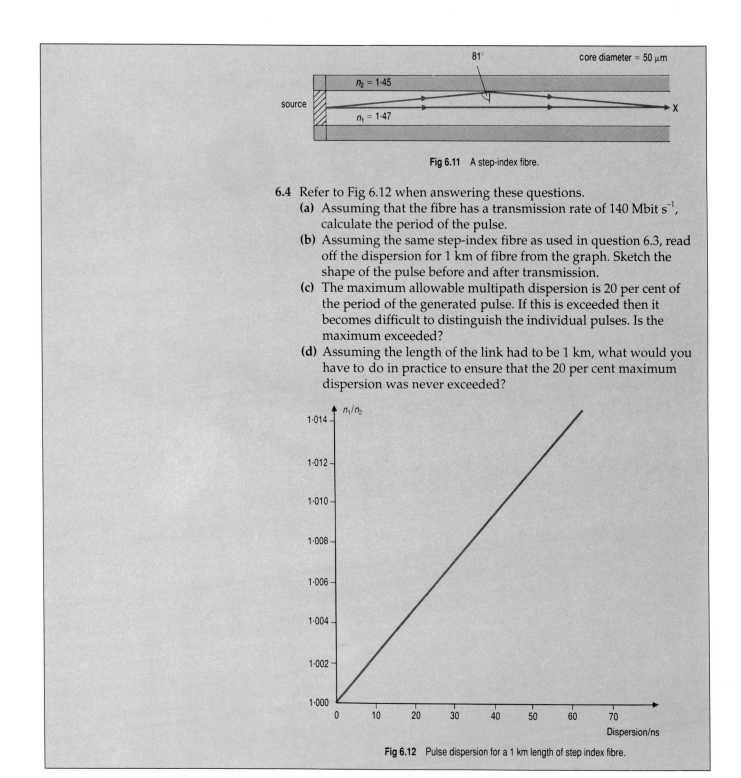

Fig 6.11 A step-index fibre.

6.4 Refer to Fig 6.12 when answering these questions.

(a) Assuming that the fibre has a transmission rate of 140 Mbit s^{-1}, calculate the period of the pulse.

(b) Assuming the same step-index fibre as used in question 6.3, read off the dispersion for 1 km of fibre from the graph. Sketch the shape of the pulse before and after transmission.

(c) The maximum allowable multipath dispersion is 20 per cent of the period of the generated pulse. If this is exceeded then it becomes difficult to distinguish the individual pulses. Is the maximum exceeded?

(d) Assuming the length of the link had to be 1 km, what would you have to do in practice to ensure that the 20 per cent maximum dispersion was never exceeded?

Fig 6.12 Pulse dispersion for a 1 km length of step index fibre.

6.3 PRODUCING AND DETECTING THE PULSES

Present digital optical fibre transmissions use PCM for encoding the signal and TDM for multiplexing the signals. The carrier is modulated using AM. The transmitter must change an electrical signal into an optical one. The ideal transmitter should emit a single wavelength, be capable of rapid pulsing and the light emitted should be coherent.

Light emitting diode

This device may be familiar to you. The **light emitting diode** (LED) is made from p- and n-type semiconductor materials similar to a normal diode but is rather more complex in structure. The commonest suitable LEDs are based on the semiconductor material gallium arsenide and they

OPTICAL FIBRE SYSTEMS

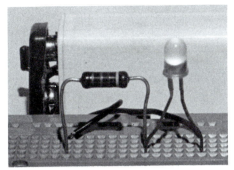

Fig 6.13 The LED.

Strictly speaking this type of modulation should really be called intensity modulating. The reason is that the current (determined by the modulating signal) through the laser of LED is proportional to the power emitted i.e. the signal intensity.

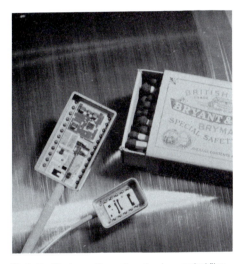

Fig 6.14 The transmitter and receiver in an optical fibre telecommunication system.

emit light when forward biased (when the anode is connected towards the positive of the power supply). It emits a small range of wavelengths and the light emitted is incoherent. The latest type of LED has a **spectral bandwidth** (the range of wavelengths emitted) of the order of 20 nm. The LED is only used with the step-index or graded-index fibres and often a lens is used to focus the light into the fibre. Although the LED does not satisfy all the criteria required for an ideal transmitter it is useful in low capacity links. Its operational lifetime is of the same order as that of a laser (about 10^5 hours) and it is a very cheap transmitter compared with the laser. Its power output is less than 1 mW which compares unfavourably with the laser's 10 mW.

Laser

The invention of the laser in 1960 led to the ideal device for producing hundreds of millions of light pulses per second. The present lasers used in optical fibre telecommunication links are based on gallium indium arsenide and are designed to produce radiation with wavelengths in the regions of 1.3 and 1.55 μm. The actual wavelength produced can be varied by altering the concentration of the doping materials during manufacture. The spectral bandwidth is about 2 nm. The light pulses are produced in a similar way to the LED – pulsing with a current, but the output is coherent. Also the emitted waves are parallel and this is very advantageous when using the monomode fibre with a core diameter of 5 μm. Fig 6.14 shows a complete transmitter (and receiver) for an optical fibre link and the size of the laser itself is roughly that of a grain of sand. Usually a glass lens is used to get the pulses into the fibre. Only the relatively high power laser can be used with the monomode fibres.

Photodiode detectors

The purpose of the detector is to change the incoming light pulses to electrical signals. The basis of the device is the **photodiode**. The photo-diode is very similar in construction to the LED, being made from p- and n-type materials, but it works the other way round. When light falls on a reverse biased diode (the anode is connected towards the negative of the power supply) a current flows that is proportional to the light intensity. There are two main types: PIN and avalanche photodiodes.

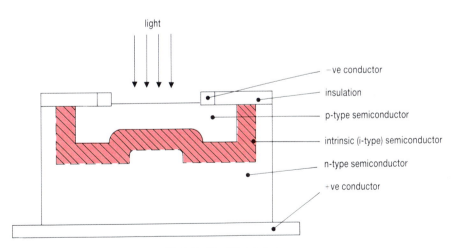

Fig 6.15 The PIN photodiode.

PIN photodiode This is the commoner type. PIN stands for positive – intrinsic – negative. This photodiode has an extra layer (intrinsic), which is lightly doped and very thin, between the p- and n-type material. The incoming radiation is detected in this area by producing electron-hole pairs. The presence of the intrinsic layer improves the sensitivity and increases the speed of operation of the detector compared with the conventional photodiode.

Avalanche diode This is similar to the PIN device but is operated at a much higher voltage. The high voltage produces ionisation and a consequent amplification of the current. The avalanche diode is more sensitive than the PIN photodiode. Besides requiring a much higher voltage for its operation its temperature must be kept constant.

INVESTIGATION

Using an optical fibre system

This investigation consists of a series of exercises to demonstrate the properties of optical fibre systems. Although the instructions are written with one system in mind other optical fibre systems could be used.

You will need: optical fibre kit (transmitter, receiver, optical fibre etc);
         signal generator;
         200 kHz pulse generator (if there is no suitable modulator with the transmitter);
         power supplies;
         double beam oscilloscope.

1. Set up the transmitter and receiver using the manufacturer's instructions.

2. Try sending signals using amplitude modulation of the light waves. The signals could be sine waves from a signal generator or sound waves using a microphone.

3. Try sending digital signals (square waves from a signal generator) along the fibre. With some systems there will be an upper limit to the frequency that can be transmitted.

4. Using various lengths of fibre it is possible to measure the attenuation rate.

5. To measure the speed of light in a fibre set up an optical fibre link similar to that shown in Fig 6.16(a). The oscilloscope (on its fastest timebase speed) will allow the input and output pulses (the pulse generator must be at least 200 kHz) to be compared directly. Use the oscilloscope (with its timebase in the calibrate position) to calculate the time taken for the pulses to pass through the fibre. Use a very short length of fibre to allow for any delay in the system and then use as long a piece of fibre as possible (at least 20 m). From this time calculate the speed of the pulses. Suggest ways of improving the accuracy of your result.

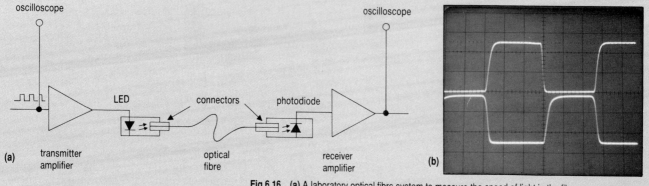

Fig 6.16 (a) A laboratory optical fibre system to measure the speed of light in the fibre.
(b) The input and output pulses obtained. The lower trace shows the input pulses, the upper trace the output pulses. The shift in position of the upper trace with the short and long lengths of optical fibre allows the transit time to be found.

OPTICAL FIBRE SYSTEMS

6.4 OPTICAL FIBRE SYSTEMS

A long distance system

The simplest type of optical fibre system was used in the last investigation. As a practical system it could only be used in a short distance and low bit rate situation. Fig 6.17 shows the sub-systems of a long distance, high capacity optical fibre link. The encoder has to change the incoming data, if analogue, to digital. The digital data is then used to modulate the optical source (a laser) turning it on for a '1' and off for a '0'. Sending 140 Mbit s^{-1} means that the laser has to be able to turn on and off 1.4×10^8 times a second. In practice, it will be considerably less than this as strings of 1s and 0s will be transmitted. The resultant series of pulses are passed into the monomode fibre.

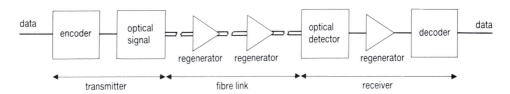

Fig 6.17 A long distance optical fibre telecommunications system.

In 1982 the then world's longest continuous optical fibre system was brought into operation between London and Birmingham as part of the telecommunications network. The world's first underwater optical fibre cable was laid in Loch Fyne in Scotland in 1980.

With long distance links it is necessary to regenerate the pulses at intervals. The regenerator requires a photodiode detector on its input side and a laser on its output side. The distance between the regenerators will be determined by the amounts of attenuation, dispersion and noise. Noise in a optical fibre system is mainly due to the random arrival of light photons and thermal noise in electronic components. At the receiver the incoming pulses are changed back into electrical digital signals before passing through the decoder. In a telecommunication system there would be a multiplexing sub-system using TDM in the transmitter section and a demultiplexing sub-system at the receiving end.

With digital systems there are a number of internationally agreed bit rates. One voice channel requires 64 kbit s^{-1}. Using TDM these can be multiplexed together to produce 32 channels with a bit rate of 2.048 Mbit s^{-1}. Only 30 of these are actually used for voice channels, the other two are used for signalling purposes such as synchronisation. Four 2.048 Mbit s^{-1} can be further multiplexed and require 8.448 Mbit s^{-1}. Further multiplexing produces 139.264 Mbit s^{-1} (or 140 Mbit s^{-1}). This is a fairly common transmission rate on the main trunk circuits but 560 Mbit s^{-1} is now coming into use.

Analogue transmissions can also take place through optical fibres. One important example of this is the transmission of cable TV where the main cables use optical fibre. Also the investigation mentioned in this chapter uses this type of modulation.

In existing systems a common transmission rate is 140 Mbit s^{-1}. In such a system it is possible to have one of the following:

The record for data transmission by early 1990 was already 1.1×10^{10} bits s^{-1} through a 260 km length of fibre without any regenerator.

- 1920 telephone circuits (bandwidth: 4 kHz and bit rate: 64 kbit s^{-1})
- 256 music channels (bandwidth: 15 kHz and bit rate: 0.5 Mbit s^{-1})
- 2 TV channels (bandwidth: 8 MHz and bit rate: 70 Mbit s^{-1}).

These can be mixed as long as the maximum bit rate is not exceeded.

Comparing optical fibre and coaxial systems

The coaxial cable was until recently the standard method for underground trunk telephone links. It is not very suitable for transmitting digital signals and the upper limit for a single coaxial cable is the order of 140 Mbit s^{-1}. Optical fibres are replacing these cables.

The advantages of optical fibres over coaxial cables are as follows:

- higher transmission capacity
- much smaller in size and weight
- the cost of fibres with equivalent capacity is less
- wider possible spacing of regenerators
- immunity to electromagnetic interference
- negligible crosstalk (signals in one channel affecting another channel)
- very suitable for digital data transmission (e.g. computer data)
- good security.

The useful properties of optical fibres described opposite have led to their being used in warships, aircraft and even motor cars in place of conventional electrical wiring. Optical digital computers have been built where the signals are carried around the computer by photons rather than electrons.

The disadvantages are:

- repair of fibres is not a simple task
- reliability of regenerators is expected to be 25 years but no one can be completely sure (this will be a problem in underwater links as it will be impossible to replace regenerators).

QUESTIONS

6.7 In an optical fibre system the signal-to-noise ratio is 20 dB and the ambient noise is 10^{-20} W. Given an input power of 10 mW, calculate:
  (a) The minimum signal power before regeneration of the pulses is necessary.
  (b) The maximum distance there could be between regenerators if the attenuation rate is 2.0 dB km^{-1}.

6.8 The following information gives some figures for an optical fibre system:
  Transmitter output: 5 mW
  Minimum receiver input: 1 µW
  Cable attenuation: 0.5 dB km^{-1}
  (a) What is the maximum attenuation, in dB, that can be allowed in the system?
  (b) Allowing for a 3 dB safety margin, what would be the maximum transmission distance possible without regeneration?
  (c) If the gain of a regenerator is 30 dB calculate the distance between regenerators in this long distance link.
  (d) In a practical system what other factors would have to be taken into consideration when determining the regenerator separation?

SUMMARY

The ability to produce glass with very low attenuation rates and sources emitting coherent light with very narrow spectral spread has produced a great step forward in telecommunications. There are three types of optical fibre in use: step-index, graded-index and monomode. The first two are multimode and only used in low capacity links. All present digital systems use PCM and TDM with AM of the carrier. With long distance systems and high transmission rates it is essential to use laser sources and monomode fibres. Optical fibre systems have many advantages over the conventional coaxial systems.

OPTICAL FIBRE SYSTEMS

QUESTIONS

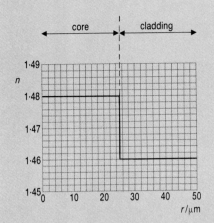

Fig 6.18

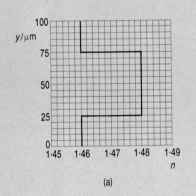

(a)

(b)

(c)

Fig 6.19

6.9 (a) Briefly discuss the ways in which light is attenuated when propagated through an optical fibre.
(b) Explain the difference between multipath and material dispersion.
(c) Explain the difference between dispersion and attenuation.

6.10 Distinguish between (i) a step-index, and (ii) a graded-index fibre, used for the transmission of optical signals. For each of these fibres explain, with the help of diagrams, how the light is propagated from one end of the fibre to the other end.

   The refractive index n of a certain optical fibre varies with distance r from the centre of the fibre as shown in Fig 6.18. Determine the critical angle for the boundary between the core and the cladding. Mark the critical angle on an appropriate diagram. (ULSEB spec.)

6.11 Fig 6.19(a) shows the variation of refractive index n with distance x across the diameter of a step-index optical fibre, which is shown in longitudinal section in Fig 6.19(b). Fig 6.19(b) also shows the passage of a meridional ray incident on the core/cladding boundary at an angle of incidence θ just greater than the critical angle θ_c.
   (i) Determine a value for the critical angle θ_c.
   (ii) If c is the speed of light in a vacuum, write down an expression for the speed of light in the core material. The optical fibre is 0.90 km long. Show that there is a difference in transit times of about 60 ns between an axial ray and a meridional ray striking B at an angle θ, just slightly greater than θ_c.
   (The speed of light in a vacuum $c = 3.00 \times 10^8 \, \text{m s}^{-1}$.)
   (iii) A microsecond pulse of monochromatic laser light whose intensity-time profile is square shaped, as in Fig 6.19(c), is launched into the fibre of A. Suggest how the pulse shape will have changed after passing through a long length of the fibre. Give reasons for your answer. (ULSEB 1988)

6.12 Describe the structure and explain the operation of a semiconductor device able to detect the signal transmitted through an optical fibre and provide an electrical output. Discuss the advantages and disadvantages of fibre optics communications compared with radio communications. (ULSEB 1987)

6.13 (a) (i) State the range of signal frequencies which defines speech bandwidth. Discuss the factors influencing the choice of frequency range, bearing in mind that the fundamental frequencies of human speech lie in a range from about 75 Hz to 300 Hz and the human ear responds to frequencies from about 15 Hz to 20 kHz.
   (ii) Information is now increasingly being carried at optical frequencies. Solid state lasers and light emitting diodes are used to convert electrical signals to light waves. Compare and constrast these two devices making reference, for example, to their power, spectral output, output directionality, maximum modulation rate, cost and lifetime. (Do not attempt to describe their structures or modes of action.)

(b) Explain the meaning of the term noise as it relates to signal transmission and state two sources of noise in the transmission and reception of radio signals.

(c) Discuss the possibility of eliminating noise at a repeater station if the signal is
  (i) amplitude modulated pulses
  (ii) binary pulses i.e. a string of equal-amplitude pulses which are either present (1) or absent (0).   (ULSEB 1988)

6.14 (a) The output signal from a microphone is in analogue form. However, there are advantages in transmitting the signal in digital form.
Explain how a digital coding represents
  (i) changes in amplitude of an analogue waveform,
  (ii) the frequency of an analogue signal.

(b) Why is the quality of speech or music transmitted in digital form generally superior to that transmitted in analogue form?

(c) Data can be transmitted in digital form through electric cable links or through glass-fibre links.
  (i) State and explain why the transit time for pulses along glass-fibre links differs from that for pulses along electric cable links of the same length.
  (ii) Explain why long distance glass-fibre links between centres of communication can be laid alongside electrified railway tracks, whereas this method is seldom used with electric cable data links.
  (iii) State briefly three further advantages of glass-fibre transmission compared with electric cable transmission.

(d) Fig 6.20 shows the waveform of digital light pulses as they enter a glass-fibre communications link.

Fig 6.20

  (i) Draw a diagram to the same scale as Fig 6.20 to show the waveform of the pulses after passing through several kilometres of glass fibre.
  (ii) Explain the consequence of not reshaping the pulses at intervals along the fibre link.   (COSSEC AS 1989)

6.15 Fibre optical systems are becoming increasingly used in modern telecommunications practice.
  (a) Identify three major advantages and one disadvantage of such a link over a radio link between two centres.
  (b) Explain why the transmission of digital signals rather than analogue signals enables a relatively 'noise-free' communications link to be established.
  (c) Why are regenerator stations needed in long distance fibre optical communications links?
  A series of optical light pulses enters an optical fibre communications link at A, as shown in Fig 6.21.

OPTICAL FIBRE SYSTEMS

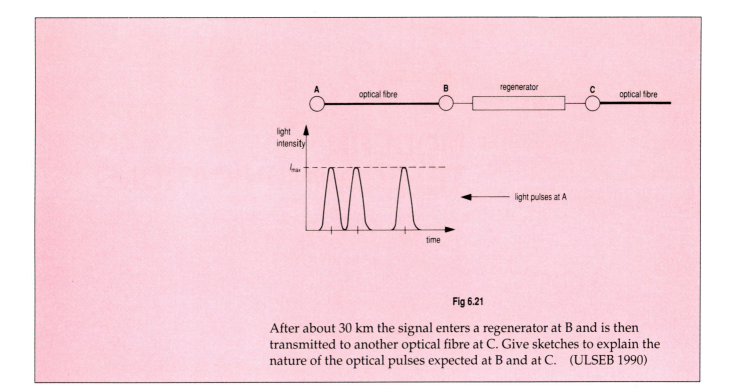

Fig 6.21

After about 30 km the signal enters a regenerator at B and is then transmitted to another optical fibre at C. Give sketches to explain the nature of the optical pulses expected at B and at C.   (ULSEB 1990)

Chapter 7

MODERN TELECOMMUNICATIONS

LEARNING OBJECTIVES

After studying this chapter you should be able to:

1. recall the factors that have led to the rapid growth and digitalisation of telecommunications;

2. recall the changes being made to the public switched telephone network and the new networks being developed;

3. describe the following developments in information transfer: telex, teletex, fax, electronic mail, teletext and viewdata;

4. describe recent and proposed developments in the following telecommunication systems: optical fibre, satellite, radio and television.

7.1 DEVELOPMENTS IN THE PUBLIC SWITCHED TELEPHONE NETWORK

Britain set for £55bn communications harvest

INNOVATION

Phones become as mobile as wristwatches

Digitised conversations bring telephones to the skies

The dream at the end of a phone

Sound, graphics, pictures – a copper telephone cable can carry them all, **Ken Young** reports

Fig 7.1 Headlines on telecommunications.

The world of telecommunications is undergoing rapid changes. This is apparent from studying the newspapers and scientific magazines such as *New Scientist*. We take it for granted that it is possible to phone direct to Australia or that the Olympics will be brought into our living room but it is not many years ago that such links were impossible. It is interesting to consider the factors that have caused such rapid growth in telecommunication facilities.

- Developments in microelectronics have resulted in greatly increased miniaturisation of the components in integrated circuits.

- The use of computers in controlling and routing signals.

- The ability to exploit higher and higher frequencies in the electromagnetic spectrum.

- Developments in signal capacity brought about by the increased use of geostationary satellites and optical fibres.

In this chapter we will look at how telecommunications are changing in the light of these developments and what will happen in the near future.

The change to digital systems

The sending of information through telecommunication links was an analogue process until microelectronics technology produced the circuitry that enabled signals to be coded digitally. A number of telecommunication links are now digital and, in the very near future, the whole system will be. There are a number of reasons for changing to digital systems:

- It is now easier and cheaper to produce digital rather than analogue circuitry.

- Digital circuitry is more reliable as it works on just two levels for the 1s and 0s.

- The effect of noise on signal propagation is much reduced. With regenerators new pulses are produced and all effects of any previously added noise removed.

- In digital circuits it is equally simple to transmit voice, music, text, graphics, pictures or computer data. Provided it has the capacity, a single channel can carry all of them at the same time.

Routing telephone calls

As mentioned in Section 1.2 the routing of calls through the **public switched telephone network** (PSTN) had originally involved the use of

Fig 7.2   Part of a system-X exchange.

electro-mechanical switching devices such as the Strowger system and later mechanical relays. By the late 1960s the complete network was automatic. Now electronic digital switching circuits are replacing this system. The exchanges being built are completely digital and, by the mid 1990s, the entire telephone system will be digital, apart from the lines into houses.

There are a number of different types of digital exchange being installed at the moment of which System X is probably the most well-known. These are computer-controlled exchanges that provide connections which are more reliable and faster than the ones in use at present. When fully operational these exchanges will provide many additional services and facilities for the user. Examples of these are:

- three way calls (three people can talk together instead of the normal two)

- call diversion (your calls can be diverted to another number)

- itemised billing of calls.

Already the telephone has undergone many changes mainly due to advances in microelectronics. Push-button keys, a memory to hold regularly used numbers, last number recall and 12-digit displays are some of the features on modern telephones. Answering machines not only take incoming calls when no one answers but allow these calls to be accessed from another telephone.

In the USA, AT and T operates probably the most sophisticated telecommunications network in the world. On the 15th January 1990, a software fault partially closed the system down for 9 hours. More than half the calls made (the 15th was a holiday and the total was below the daily average of 100 million) were blocked because of this fault.

In the near future it is possible that each individual requiring a telephone will be allocated his/her own personal number. The telephone itself will be a small radiophone that can be carried around. The demand for telephone numbers in the UK will increase from 3×10^7 (1989) to 4×10^8 by the middle of the next century.

Fig 7.3 A modern telephone.

It is cheaper for two computers to 'speak' to each other across the Atlantic Ocean via the PSS lines than it is for two people to converse with each other via the normal telephone circuits.

New networks

The PSTN has used coaxial cable in its trunk (i.e. long distance) routes for many years but this is now being replaced by optical fibre. In 1989 there was about 400 000 km of optical fibre in the network. Signals are digital and the fibres can then carry voice, music, television, pictures or computer data in digital form as required. In the latest systems a single optical fibre can carry 560 Mbit s^{-1}, though this will undoubtably increase in the near future.

In addition to changes in the main network there are a number of additional networks being laid to cater for today's need for high speed high capacity data transmission links. A new network that is laid in parallel with an existing system is called an overlay. Until there is complete integration of analogue and digital signals (see the next section), British Telecom (BT) has provided links just for digital transmission. These links are called Kilostream (maximum rate is 64 kbit s^{-1}) and Megastream (maximum rate is 2 Mbit s^{-1}). Kilostream provides a single channel for data, speech, high speed fax or slow-scan TV while Megastream allows multiple channels. These are not switchable and are private leased lines being laid from one fixed point to another according to the customer's requirements.

Packet Switch Stream (PSS) is a switched network like the PSTN. It was introduced in the mid 1970s to transfer digital data. The data is 'packeted' and sent by time division multiplexing techniques so that several separate messages can be handled at the same time over one line. Special interfacing equipment is required to link the user into this system. International PSS is available to some 80 countries.

Since the privatisation of BT there has been an opportunity for other companies to work in telecommunications. To date only one company, Mercury Communications, has installed its own network though, in many cases, it has to link into BT's PSTN. Mercury is aimed mainly at business users and has installed a trunk network of optical fibre. This covers England in a figure-of-eight loop and it is linked into cities at various points. Most of its network uses digital transmission. Mercury's users have access to satellite communications and the other types of facility available to the BT user.

7.2 THE TRANSFER OF INFORMATION

Information can be in a number of different forms. Apart from sound it can be text, pictures or data. There are ways of sending all of these via a telecommunications channel. They can be sent either as analogue signals using the existing PSTN, via the newly-laid digital links or using radio waves. Nowadays the transfer of information quickly and accurately has become essential to a large number of organisations and individuals.

Telex

Telex started in 1932 as a telegraph service. There was a terminal with a device similar to a typewriter called a teleprinter for each user. The number of the receiver was dialled and, when contact was established, the message could be typed and sent simultaneously or a previously encoded punched tape could be used. The system has advantages over the telephone when it is essential to send written documentation. It became automatic in 1958 and is the largest public messaging network in the UK. The modern telex machines have electronic memories and word processing facilities. In 1989 there were over 115 000 users in the UK and more than two million users in over 200 countries. Although telex machines are sophisticated this service may be superseded by methods such as teletex (see the next paragraph). But the telex network has its own lines and the size of the international telex network will probably give it many more years of life.

Fig 7.4 Telex terminal.

Teletex

This is the modern equivalent of telex. It is a high speed message transmission system which can send a typical A4 page of information in about 30 seconds. It was opened in 1985 and operates over the telephone or PSS networks. The message can be prepared on the teletex terminal using word processing facilities and then transmitted in a similar way to telex. Any electronic text equipment can be used if a suitable teletex interface is provided.

Fax

Facsimile transmission (fax) is the name given to the transmission of documents and drawings over the national and international networks. Exact copies of a document (graphics or text) can be sent between two facsimile machines in less than a minute on the PSTN (with digital links it is faster than this). It is rather like having a photocopier that produces the copy a long distance away. In 1988 there were approximately 250 000 UK fax users and about 100 countries to which documents could be faxed.

Fig 7.5   Fax machine.

Electronic mail

Traditionally telex has been the main method of electronic text communication together with the more sophisticated fax and teletex systems. It is now possible to integrate electronic messaging with office automation functions for fast and flexible business communications. This system allows users to communicate via the PSTN using an 'electronic mailbox'. Each user has a computer together with a modem and the appropriate software. The **modem** (**mo**dulator/**dem**odulator) changes the binary digits from the computer into analogue signals more suitable for transmission along the PSTN lines. The 1s and 0s are changed into signals of two different frequencies. At the receiving end the process is reversed to obtain the original data. The sender composes the message on the computer and then transmits it with added codes to indicate the address of the recipient's mailbox. The recipient can access the system's mailbox and download to his/her own computer.

There are a number of electronic mail systems in existence. BT's system is called Telecom Gold and Mercury's is called Link 7500. Apart from the cost of using the phone lines users have to pay for the service. Users have access to the worldwide telex network giving them access to more than 200 countries for the transmission of text to colleagues and customers at any time.

Teletext

This system allows text that is broadcast with the television signals to be displayed on the screen. The carrier of the TV signal has 'spaces' on it which can be used to carry information that is displayed on the screen. It requires a special decoding unit in the TV and a keypad to access the information (this is the only cost to the user). The pages of text are sent in

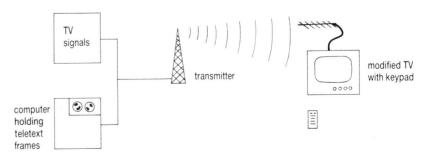

Fig 7.6   The transmission of teletext.

sequence and this causes the delay often experienced when selecting a particular page. There are two systems in operation in the UK – CEEFAX (BBC) and ORACLE (IBA).

Viewdata

Unlike the teletext services where information only travels one way (non-interactive), viewdata allows for the two-way (interactive) passage of information. The first public viewdata system in the world was **Prestel** which was started in 1979. The 'pages' of information (at the moment there are about 250 000 of them) are stored in a number of computers around the

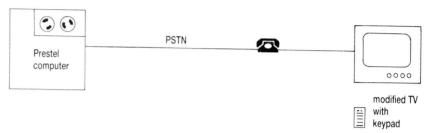

Fig 7.7 The transmission of viewdata.

country. The information is accessed via the PSTN and can be displayed on a specially adapted TV set with the help of a remote control keypad. By keying in the required number the user can be connected to the nearest Prestel computer and request the required pages. Also information can be sent back by the user to the information provider via the Prestel computer. It is not a free service and the user is billed for cost of telephone calls and access to the pages used, though some pages are free. Many people use their computers to access Prestel and other viewdata systems or databases. The computer has to be linked to the telephone line via a modem.

Prestel Gateway extends the viewdata system to include a large number of other computer networks. It is possible to access these systems for extra services or information. For example, you could book an airline flight directly or order a new piece of furniture. At the moment only text and low resolution sketches can be distributed via the PSTN but with the coming of the completely digital network high quality pictures will be displayed on the pages.

Fig 7.8 A page from Prestel.

Integrated Services Digital Network

British Telecom has three separate switched networks which provide different telecommunication services:

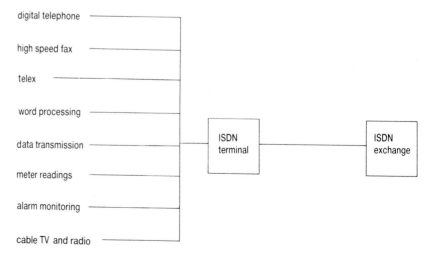

Fig 7.9 The ISDN system.

- the public switched telephone network (PSTN)
- telex
- packet switch stream (PSS).

The aim of this system, usually abbreviated to ISDN, is to provide a single network that will carry all of the existing services and any additional ones that will be required in the future along a single digital link. Some of these are illustrated in Fig 7.9 but these are just some of the possibilities. It will be a few years before the ISDN is completed as there are many problems to overcome, not the least being the incompatability of the existing systems in different countries (see Fig 7.10). However, the European Parliament has decreed that all its member countries shall have 10 per cent of their PSTN capable of carrying ISDN services by 1992.

Seventeen years ago plans were devised for a worldwide digital network linking voice, data and image. Today the Integrated Services Digital Network (ISDN) is the dream becoming reality. The world's leading telephone suppliers are creating digital networks that enable the existing copper cabling to carry a combination of sound, printed words or graphics from potentially any electronic device.

The benefits are many. First, it means that hitherto separate technologies can be combined in innovative ways: the personal computer can have an integral telephone or video TV link; a caller can transfer and discuss a file of computer data without breaking the call.

Second, agreed standards being developed could make ISDN into a global network in much the same way as the existing telephone system.

Because ISDN offers fast data transfer, a range of digital services will come into use – high speed fax, fast text transfer, digital phone services, video-conferencing.

A standardized agreed form of ISDN is being developed that will enable networks across Europe and even worldwide to be inter-connected. In the UK, British Telecom is proud of having introduced the world's first commercial non-standard ISDN service, dubbed single line Integrated Digital Access (IDA), in 1985.

Despite possible benefits, the take-up in Britain has so far been slow, perhaps because the IDA service is non-standard and arrived later than expected; many users are confused and sceptical.

In Europe, France and West Germany are expected to have nationwide services in place by the end of this year. France Telecom plans to have 150,000 users by 1992 and France and West Germany expect to link by 1990.

The Times 2 June

Fig 7.10   A newspaper article on the proposed European ISDN.

7.3 DEVELOPMENTS IN TELECOMMUNICATION SYSTEMS

Satellite systems

As the number of geostationary satellites increases so does their number of uses. Originally they were used for transferring telephone and television signals between continents or widely spaced countries but they have now been extended into many other areas. Examples of these are:

- data transmission
- videoconferencing (allowing groups of people, perhaps in different continents, to see and communicate with each other)
- publication of newspapers (the newspaper can be written in one country, transmitted and then printed in another country)
- educational purposes (a project in India beams educational material from a satellite down to remote villages and towns)
- direct broadcasting by satellite (DBS) (see next paragraph)

14/11 GHz band   The use of frequencies in the 14/11 GHz band means that smaller dish aerials can be used. It is possible to provide temporary links for important events (many outside broadcasts are initially beamed up to a satellite nowadays) or a business can fix the aerial outside their building and use satellite channels on either a temporary or permanent basis. With smaller aerials it has become possible to transmit television

Fig 7.11   A transportable satellite terminal.

signals to individual homes. The frequencies are not the same as for telecommunications being in the 14/12 GHz range. In 1977 each country in Europe was allocated five channels in the 12 GHz range for DBS. In 1989 the first DBS system in the UK using these frequencies started broadcasting. Power requirements are much higher than on telecommunication satellites and this system uses the satellite called ASTRA which broadcasts at 45 W per channel.

INTELSAT The telephone traffic over the INTELSAT satellites is declining in the face of competition from cable systems. This can in part be offset by concentrating on rural telephony and services for the developing world, using small aperture antennas. Digital techniques have been developed to increase the trunking capacity with an expected reduction in the cost per call. One of these techniques is time division multiple access (TDMA) which operates at 120 Mbit s^{-1} and each earth station transmits its traffic in a 2 ms burst. INTELSAT VI satellites feature an on-board switching capability to give dynamic routing of transmissions from various uplinks to specific downlinks. Developments in the technology have greatly increased the capacity of satellites. With INTELSAT VI the original 30 000 voice channels can be increased to 132 000 by using different types of signal polarisation and 'spot' beams.

EUTELSAT The European Telecommunications Satellite Organisation (EUTELSAT) deals with Europe's satellite telecommunications and has 25 members. EUTELSAT first launched a telecommunications satellite in 1982 with 12 000 voice and 2 television channels. The latest of their five first generation satellites was launched in 1988. Due to their close links with the European Broadcasting Union much of their capacity is used for transmitting television (in the 14/12 GHz band) but telephone traffic has increased steadily (in the 14/11 GHz band). The second generation of EUTELSAT satellites will start to be launched in 1990.

INMARSAT Radio communications between ships and between ships and shore have never been simple due to the distances involved and the variability in transmitting conditions. A satellite system was set up to overcome these difficulties in 1982 by the International Maritime Satellite Organisation. It consists of three geostationary satellites placed above the Atlantic, Pacific and Indian Oceans. It now supports maritime, aeronautical, land mobile and international satellite paging services. The aeronautical service started in 1989 and is called Skyphone. This service is available to aeroplanes and airlines providing voice and data facilities by direct dialling.

Optical fibre systems

In addition to their replacement of coaxial cables, optical fibre networks are being built to cater for the rapid increase in demand for specialised communication links within the business community. The first of these is called the flexible access system (FAS) and is being installed in London. The first stage opened in 1988. Optical fibre goes into the customer's premises and will allow access to a wide range of services. The first phase will use an estimated 60 000 km of fibre. Later it will be extended to other parts of London and other cities in the UK. FAS will link directly with the growing ISDN.

TAT–8 This is the name for the transatlantic optical fibre submarine cable that became operational in December 1989. It is capable of carrying 40 000 simultaneous channels. The main cable consists of six fibres, two pairs carry the traffic and the third is used for back-up. The repeaters are spaced

It is estimated that by 1996 there will be the following underwater oceanic optical fibre systems: six trans-atlantic networks, three transpacific networks and two systems between Australia and Hawaii. Also, it is estimated that the total investment, including shorter underwater links, will be of the order of £4 billion.

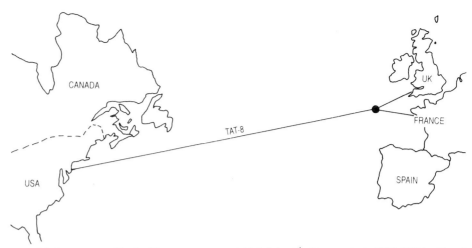

70 km apart. Each fibre can carry 140 Mbit s^{-1}. One pair is UK–USA direct, one pair UK–France and the other USA–France. It is buried in the sea bed at a depth of 1 m until the depth is 1 km and is steel clad to a depth of 2.6 km. The cost of the cable was £220 million.

TAT–8 is basically capable of carrying almost 8000 voice channels (i.e. 4000 telephone conversations) of 64 kbit s^{-1} but each channel can carry up to five voice channels using what is known as digital circuit multiplication equipment. This equipment can be used to increase the capacity of any digital system, provided that it is telephone signals rather than digital data that is transmitted. In 1991 it is expected that a complete optical fibre link will be in service between the UK and Japan via Hawaii.

Radio and television

Broadcasting   There will probably be little change in the technology involved with radio and television but there will be an increasingly wide range of choice of programmes as government regulations are changed. High definition television (HDTV) sets with 1250 lines (rather than 625) will soon appear. The quality of the picture is greatly improved but at the expense of a slight increase in transmission bandwidth.

Manufacturers hope that High Definition TV (HDTV) will be operational in time for the 1992 Olympics in Barcelona. The HDTV screen will have 1250 lines each made up from 1440 pixels (at present a TV screen consists of 625 lines and 720 pixels). Also the HDTV screen may be up to 130 cm across. The HDTV picture will contain four times the information content of present screens but will use a similar bandwidth.

Cellphone systems   The latest version of this started in 1985 using frequencies just below 1 GHz and the majority of users have the cellphones in their cars. There are two non-compatible systems in existence – Cellnet and Vodaphone. The country is divided up into 'cells' each of which has a low powered transmitter/receiver at its centre. These are joined via electronic exchanges into the PSTN. Although usually drawn as hexagons, the cells are actually circular and range in diameter from 2 km in cities where there are a large number of users to 30 km in the country. There are restrictions on the number of channels used (300 for each company). Each pair of users is allocated one of the available channels when they make a connection and so there are obviously limits to the number of callers that can be on the air in a particular cell at the same time. As the user travels from one cell into another the signals are automatically transferred to the new cell's transmitter/receiver.

Radiopaging is a system designed for people who are away from their base or difficult to reach. The caller telephones the paging number and the device 'bleeps' automatically. At present the system covers most of the UK but there will a paging service covering the whole of Europe by 1993.

Cordless telephones   This new generation of cordless telephones, unlike those used at present, allows the user to link into public base stations. The system is called Telepoint and is completely digital. The accessed station then links the caller into the public networks. Each handset has a unique identification code and a central computer logs the call and charges it to the appropriate account. The target is the establishment of 95 000 base stations each capable of receiving calls from a cordless telephone up to 200 m away. The power requirements for the handset will be much smaller than that for a cellphone. However the calls using this system are only one way.

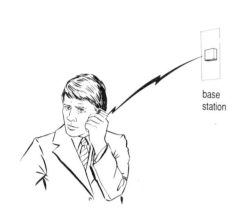

Fig 7.14 The cordless telephone.

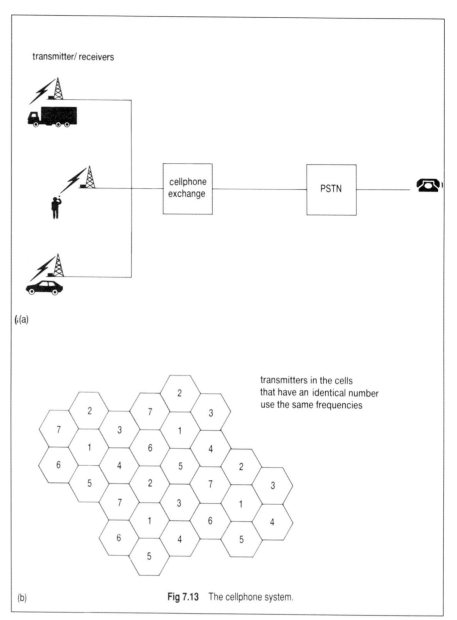

transmitter/ receivers

cellphone exchange

PSTN

(a)

transmitters in the cells that have an identical number use the same frequencies

(b)   Fig 7.13 The cellphone system.

In such a rapidly developing area as telecommunications predicting what will happen in the future is a very speculative task. The present rapid growth will probably not be sustained and the growth of international communications will probably not increase as fast as the national networks. In the UK the completion of the ISDN network will allow each individual access to all available telecommunication facilities. Whether individuals will be willing to pay for all the sophistication is rather debatable. The development of the ISDN with an optical fibre link into the home will allow many facilities for the individual. Examples are:

- a wide range of home entertainments from many sources including HDTV
- access to a wide range of educational material
- access to remote databases
- access to interactive data and information services
- instant electronic mail
- videophones (visual and sound communication)
- shopping and banking from home
- remote meter reading for electricity and gas
- alarm systems for fire, burglary and for elderly or handicapped people.

It has been suggested that the coming of such facilities will increase decentralisation of offices and businesses. Access to all telecommunication facilities means that an office could be situated in the country just as conveniently as in a town or city. Also many people could just as easily work from home as commute to an office.

Whatever happens the place of telecommunications in the present information technology revolution is of paramount importance. How we use the increase in facilities will determine many of the changes, both good and bad, that will occur in our society.

New satellite services

With the coming of small receiving aerials satellite services will become very competitive especially where terrestrial microwave networks are not available or cost effective. The receiving systems are called very small aperture antenna terminals (VSAT). The satellite capacity is now available and will be increased when the EUTELSAT II satellites are launched from 1990 on. It is estimated that by 1994 as much as 40 per cent of EUTELSAT's capacity will be used for VSAT and that there will be more than 40 000 such terminals. The services provided will contain video, voice and high quality data links. However ISDN will be a strong competitor to this system.

New spacecraft technologies

In 1992 the first of the INTELSAT VII series is expected to be launched. These are designs for the Pacific Ocean region. In the past satellites were designed for the more demanding Atlantic region. INTELSAT VI uses the 6/4 GHz band six times while INTELSAT VII will use it only four times. This reduction is intentional in order to reduce costs and match the satellite technology to its operational roles.

In general the introduction of new technologies must always be cost effective and controlled by the evolution of other traffic requirements. On-board processing, inter-satellite links, reconfigerable satellite antennas and new methods of spacecraft propulsion are all being developed at the moment. The 30/20 GHz bands will come into use in the near future.

Terrestrial microwave systems

Traditionally the 2–11 GHz band has been the most commonly used but the demand is dropping for high capacity digital links on the major trunk

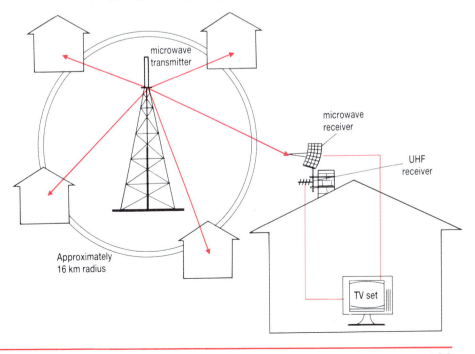

Fig 7.15 Multipoint microwave distribution.

routes as optical fibre systems are installed. However, short distance local links at higher frequencies are being installed as an alternative to optical fibres. One possible future use is in the development of multipoint microwave distribution (MMD) which would distribute 30 TV channels to homes at 25 GHz over a radius of 16 km, providing an alternative to satellite DBS systems.

Pan European cellular radio

The existing European systems are not compatible and a new system has been designed for the 1990's. A digital voice system will be used at 16 kbits^{-1} or less. A user will be allocated a time slot in which to transmit rather than a frequency slot as at present. It is estimated that there will be 10 million users by the end of the 1990s.

Optical fibre systems

Silica based optical fibres are coming into use with attenuation rates of 0.15 dB km^{-1}. The newer fluoride glasses suffer less from Rayleigh scattering than silica based ones but losses are at present too high. However it is expected that this loss can be reduced to 0.005 dB km^{-1} at 2.1 μm. Wave division multiplexing enables a number of transmissions at different wavelengths to be sent along the same optical fibre. The use of optical switches allows the signals to be multiplexed together for transmission and demultiplexed at the receiver. Repeaters are being developed that contain laser amplifiers. In the not too distant future these improvements will produce optical fibre systems with 200 km between repeaters capable of transmission of 10 000 channels.

Among the many improvements being developed for the car radio is one that will translate foreign traffic reports. Traffic reports will use only a limited number of words each with its own digital code. If you are driving in Europe then any traffic report can be instantly translated via a speech-synthesiser chip in the radio.

SUMMARY

The development of high capacity digital links is producing rapid change in telecommunication systems. With the help of completely digital exchanges the PSTN will be digital in a few years. The development of telex, teletex, fax and electronic mail have produced increased telecommunication facilities. Teletext and viewdata have produced information services accessible in one's home or office. Optical fibre and satellite systems are taking over an increasingly larger fraction of the telecommunication links. The production of the ISDN will allow all telecommunication facilities to be transmitted on the same network.

Appendix A:
ANSWERS

Chapter 2

2.5 **(a)** 10^8 Hz
(b) 6 m
(c) 2×10^5 Hz
(d) 2.8×10^2 m

2.6 **(b)** 125; 1.5×10^4

2.7 **(a)** 6
(b) 48 kbit s^{-1}
(c) 56 kbit s^{-1}

2.8 **(a)** 10^{-9} s
(b) 10^{-9} s
(c) 6.4×10^{-5} s

2.9 **(a)** 7 μs
(b) 125 μs
(c) 17

2.10 125

2.11 **(a)** 10^{-1} W
(b) 10^{-1} W

2.12 **(a)** 1.3×10^{-18} W
(b) 62.3 km

2.14 **(a)** 12 Mbit s^{-1}
(b) 96 Mbit s^{-1}
(c) 96 MHz

2.15 **(a)** 3(11), 5(101), 7(111), 7, 7, 4(100), 2(10), 0

2.16 **(c)** 1.2 kHz
(d) 4×10^4

2.17 **(a)** − 60 dB
(b) 10^{-9} W

Chapter 3

3.2 **(a)** 3.4 m, 2.8 m
(b) (i) 270 (ii) 2.7×10^4

3.5 **(a)** 120
(b) 1.0045 MHz, 0.9955 MHz
(c) 1048.5 to 1052.95, 1053, 1053.05 to 1057.5 kHz, 9kHz

3.13 **(b)** (i) 6.8 kHz
(ii) 196.6 to 199.7 kHZ
(iii) 200.3 to 203.4 kHZ

Chapter 4

4.2 **(a)** 628 Ω , 25.4 kHz
(b) 4.1 kHz
(c) 250 nF
(d) 5.1 mH
(e) 1.6 MHz

4.4 **(a)** 49 μH

4.5 **(d)** 2 MHz, 4 MHz
(e) 6.3

4.10 **(a)** 40.8 kHz
(c) 43.9

4.11 **(b)** 69 μH

Chapter 5

5.1 **(b)** 15

5.2 **(a)** 95 min
(b) 3.1 km s^{-1}
(c) 84 min, 7.9 km s^{-1}

5.6 **(a)** 49 dB
(b) 0.3°

5.7 **(a)** 40, 36 dB
(b) 176 dB
(c) 0.4 W

5.8 **(a)** (i) 60 dB
(ii) 0.35°
(iii) 0.18°
(b) 1.1×10^5 m
(c) 2.4×10^3

5.9 **(b)** 4.7×10^{-2} m

5.10 **(a)** (ii) 2.0 m
(b) (i) 2.998×10^8 m s^{-1} 2.998×10^{-2} m
(d) (i) 0.5 P $(1 + 10^{-5})$
(ii) 5×10^4 (47 dB)

5.11 **(b)** (i) 0.24 s
(ii) 0.03 s

Chapter 6

6.1 **(a)** 4.61 km^{-1}
  (b) 10^{-12} W
  (c) 100

6.2 **(b)** 2187
  (c) 1

6.3 **(a)** $80.5°$
  (b) $2.04 \times 10^{8} \text{ m s}^{-1}$
  (c) 4.0 μm
  (d) $2.0 \times 10^{-14} \text{ s}$

6.4 **(a)** 7.14 ns
  (b) 60 ns

6.7 **(a)** 10^{-18} W
  (b) 80 km

6.8 **(a)** 37 dB
  (b) 68 km
  (c) 54 km

6.10 $80.6°$

6.11 (i) $80.6°$
  (ii) $2.03 \times 10^{8} \text{ ms}^{-1}$

Appendix B:
FURTHER RESOURCES

FURTHER READING

General reading

This section refers to books that would provide useful supplementary information at a number of places throughout the text.

Telecommunications in Practice General editors Brian Nicholl and Jenny Selfe (Association for Science Education/British Telecom)
Contains a wide ranging series of articles written by teachers covering most areas of the telecommunication scene. A very useful book.

Telecommunications Primer by Graham Langley (Pitman)
Covers the basics of telecommunication systems. Contains sections on telephones, cables (submarine and optical), radio, television and satellites. Two chapters deal with digital aspects of telecommunications and there is a detailed look at the future of telecommunications.

Telecommunications Systems by P. H. Smale (Pitman)
Provides a simple introduction to the principles behind telecommunication systems. There are sections on radio, television, telephones and data transmission.

The Penguin Dictionary of Telecommunications by John Graham (Penguin Books)
A detailed guide to the terms used in telecommunications today. The explanations are written in a simple form and are cross-referenced to other terms.

Electronic Communication Systems by Frank Dungan (Breton Publishers)
Covers a wider area than required. However, it is very comprehensible and a useful resource book.

Chapter References

This section refers to books that would provide useful supplementary information in the appropriate chapters.

Chapter 1

Broadcasting in Britain 1922–1972 by Keith Geddes (Her Majesty's Stationery Office)

Names and Dates for Students (British Telecom)

Pioneers in Telecommunications (British Telecom)

Radio and British Telecom (Her Majesty's Stationery Office)

Submarine Telegraphy the Grand Victorian Technology by Bernard S. Finn (Her Majesty's Stationery Office)

Telecommunications – A Technology for Change by Eryl Davies (Her Majesty's Stationery Office)

Television – The First 50 Years by Keith Geddes and Gordon Bussey (Philips Electronics/National Museum of Photography, Film and TV)

Chapter 2

Reliable Digital Communication (Hobsons)

Telecommunications Transmission Systems (Hobsons)

Transmission Systems II by D.C. Green (Pitman)

Chapter 5

UoSAT: The University of Surrey Satellite Project (Department of Electronic and Electrical Engineering, University of Surrey)

Satellite Dynamics (Hobsons)

Chapter 6

Communication and Light (Hobsons)

Chapter 7

Beyond the Telephone: The Intelligent Network (British Telecom)

Getting the Message (British Telecom)

RESOURCES

Chapter 2

Software: *Signal Transmission*
The program allows the user to investigate a number of factors that affect signal transmission: attenuation, amplification, noise and distortion. The differences between analogue and digital can also be studied. Although a telephone channel is used in the program the principles are applicable to other telecommunication systems.

Pulse Code Modulation
This program allows the investigation of encoding analogue signals using PCM. The effects of varying the number of sampling levels and the sampling rate can be studied.

Both programs available from: Computers in the Curriculum Project for computers: RML 380Z, 480Z, Nimbus, stand-alone and network BBC B, Master, stand-alone and network.

Telecommunications in Practice: Section 19 contains instructions for setting up a TDM link using two channels.

Investigation: *Synthesis of waveforms from sine waves*
Suitable software showing the addition of sine waves is available from AVP and is called *Adding Waves*. It is available for most of the computers used in schools and colleges.

Chapter 3

Investigation: *A simple transmitter.*
As a further investigation, the construction of a spark transmitter and coherer receiver is described in Physics Education, vol. 24 no. 5 p. 312–3.

Investigation: *AM/FM of a carrier wave*
Suitable signal generators with AM/FM inputs are produced by Philip Harris (P63710/0) and Unilab (062.101).

Investigation: *The wave properties of radio waves*
Suitable apparatus is available from Unilab: microwave transmitter (044.672) and receiver (045.674) together with the relevant accessories mentioned in the investigation. It is possible to modulate the transmitted microwaves by using an unsmoothed power supply and this can be detected (100 Hz) by using an audio amplifier connected to the receiver.

Chapter 4

Investigation: *'Looking at' r.f. transmissions*
A suitable tuning circuit module is available from Unilab (223.020).

Radios:
Apart from the Alpha Kit radio manufactured by Unilab (units are 223.020, 220 and 221) illustrated in Fig. 4.16, there are two other radios made by Unilab that could be used for construction of an MF band radio. Both are based on the ZN 414 chip:
The Abergele Communications I Board (540.400). A sophisticated circuit board that has many other uses in the area of communications.
The Nuffield Co-ordinated Sciences radio kit (093.020 and 021). It comes in packs of five and is a simple kit to assemble. Tuning is accomplished by varying the inductance.
ZN414Z and ZN414E are available from RS Ltd together with Datasheet 5859 on their use.

Investigation: *Testing a dipole aerial*
Suitable apparatus is available from Unilab (063.661). The transmitted signal is modulated at 400 Hz and this can be detected using an audio amplifier. A spark transmitter is included in the kit – it will need a 5 kV supply and a light beam galvanometer or d.c. amplifier for detecting the received signal.
Operational amplifier: A suitable module is available from Unilab (511.026).

Chapter 5

Accessing satellites:
The following systems have been designed for educational use. All relevant details are given in the documentation accompanying the equipment.

Available from Griffin and George: Microsat Weather Satellite Receiving Systems for accessing METEOSAT and NOAA satellites. The system can be linked to the following computers: BBC B and Master, Archimedes, Apple II GS, IBM PS/2 model 50 and Nimbus.
Available from Philip Harris: 'Astrid' Satellite Receiver System for

accessing UoSAT satellites. The receiving system can be linked to BBC machines or a Spectrum.

Available from Unilab: UoSAT Ground Station Package (532.161) NOAA Weather Satellite Ground Station Package (532.160) Combined NOAA and UoSAT Ground Station Package (532.162) Each of these packages provide complete ground stations to be set up (apart from the BBC B or Master plus monitor).

Dartcom System for accessing METEOSAT or polar orbiting satellites broadcasting between 136 and 138 MHz. The receiving system can be linked to Nimbus or Apple Macintosh II.

Video: *Satellite Communications*
Available from British Telecom: This provides a useful introduction to the principles behind satellite telecommunications. Two booklets accompany the video.

Chapter 6

Telecommunications in Practice: Section 9 contains useful information on the making and testing of optical fibres

Optical Fibres in School Physics (No 2 in the series Experimenting with Industry published by the ASE): This contains a wealth of information on the practical aspects of using optical fibres.

Video: *The Physics of Optical Communications*
Available from: British Telecom This provides a useful introduction to the principles behind optical fibre telecomunications. Two booklets accompany the video.

Investigation: *Using an optical fibre system*
Suitable apparatus is available from Philip Harris (P42000/7) and is shown in Fig 6.17. Other suitable systems are available from Unilab (223.231 and 232) and from RS Ltd (302-081)

PLACES TO VISIT

Science Museum, Exhibition Road, London, SW17 2DD. Phone: 071 938 8111. There is a Telecommunications Gallery on the first floor that provides a very detailed history of the subject.

Telecom Technology Showcase, 135 Victoria Street, London EC4V 4AT. Phone: 01248 7444
Shows the development of telecommunications from the beginning up to the present. Besides the exhibits there are videos that can be viewed showing the latest developments in telecommunications.

Goonhilly Satellite Earth Station, Helston, Cornwall. Phone: Helston (0326) 574141
Has a visitors centre that describes the history of Goonhilly, its present work and future plans. There is also a coach tour and audio-visual display.

Madley Satellite Earth Station, Madley, Herefordshire. Phone: Madley (0981) 250001
Similar to Goonhilly.

Index